Materials for a Sustainable Tomorrow:

Bridging Science and Society

Materials for a Sustainable Tomorrow: Bridging Science and Society

Editors

Dr. Leeladhar Chouhan

Assistant Professor

Department of Physics, Kokrajhar University

Dr. Rimpi Chakravarty

Institute Post Doctoral Fellow

Department of Physics, Indian Institute of Technology Guwahati

ALLIED PUBLISHERS PVT. LTD.

Noida • Mumbai • Kolkata • Chennai

ALLIED PUBLISHERS PRIVATE LIMITED

D-5, Sector-2, **Noida**–201 301
Ph. Nos.: 0120-4320295/2542557/4352866 • E-mail: delhi.books@alliedpublishers.com

17 Chittaranjan Avenue, **Kolkata**–700072
Ph.: 033-22129618 • E-mail: cal.books@alliedpublishers.com

15 J.N. Heredia Marg, Ballard Estate, **Mumbai**–400001
Ph.: 022-42126969 • E-mail: mumbai.books@alliedpublishers.com

No. 25/10, Commander-in-Chief Road, Ethiraj Lane (Next to Post Office)
Egmore, **Chennai**–600008
Ph.: 044-28223938 • E-mail: chennai.books@alliedpublishers.com

Website: https://alliedpublishers.in

ISBN: 978-93-48699-57-2

Published by Sunil Sachdev and printed by Ravi Sachdev at Allied Publishers Pvt. Ltd.,
D-5, Sector-2, Noida–201 301

Editorial Advisory Board

Editorial

As we grapple with the escalating realities of climate change, resource scarcity, and environmental degradation, the materials that form the bedrock of our civilization must undergo a radical evolution. It is with this sense of urgency and purpose; we aimed to bring out this book '*Materials for a Sustainable Tomorrow: Bridging Science and Society*'.

This book was born out of the necessity to bridge the gap between high-level laboratory research and its practical, societal application. While scientific advancements in green chemistry and nanotechnology are flourishing, their success depends on interdisciplinary collaboration, supportive policy frameworks, and public literacy. This volume serves as a roadmap for that integration, providing a holistic understanding of how sustainable materials can address pressing global challenges.

The 17 chapters compiled in this volume are not merely a collection of studies; they represent a 'logical intelligence-driven flow' that mirrors a Technology Readiness Level (TRL) progression. By structuring the chapters from raw synthesis to frontier technologies, we provide a narrative of how sustainability matures from an idea into a societal pillar: green synthesis → energy production → storage → environment → structures → frontier tech.

A project of this magnitude is the result of collective dedication. We wish to thank our Advisory Board and the contributing authors. Your rigorous scholarship and visionary thinking have ensured that this book is not just a publication, but a catalyst for change. We also acknowledge all the reviewers for their comments and efforts which helped to improve the quality of the chapters. We extend our gratitude to Allied Publishers Pvt. Ltd, for their unwavering support and commitment to bringing this volume in its stipulated time.

Date: 10-02-2026 **Editors**

Contents

CHAPTER 1

Green Pathways for Nanomaterial Synthesis: From Nature to Nanotechnology

Rajnandan Lahkar, Hrishikesh Sarma, Ipshita Baruah and Hemen Kalita*

[1]Nanomaterials and Nanoelectronics Laboratory, Department of Physics, Gauhati University, Assam, India

*E-mail: hemenkalita@gauhati.ac.in

ABSTRACT: *Nanomaterials such as graphene, transition metal dichalcogenides (TMDs), layered metal dichalogenides (LMDs), hexagonal boron nitride (h-BN), and MXenes are intriguing materials as they exhibit exceptional electronic, mechanical, and thermal properties, which make them important in various applications such as electronics, energy storage, catalysis, and sensing. However, traditional methods to synthesize these materials often utilize toxic chemicals, high synthesis temperatures, and energy-intensive processes that affect environmental sustainability and occupational safety. Green synthesis pathways using natural precursors and eco-friendly solvents for nanomaterials offer an environmentally friendly and inexpensive alternative. In this chapter, we discuss various traditional methods to synthesize nanomaterials, and in the latter part, we introduce a general overview of sustainable methods to synthesize nanomaterials using eco-friendly green approaches. Thus, nanotechnology, combined with green pathways, can help decrease the carbon footprint of making materials while producing scalable and consistent outcomes.*

Keywords: Nanomaterials, Green Synthesis, Traditional Synthesis, Sustainable Nanotechnology, Eco-friendly Methods, Biocompatible Nanomaterials, Environmental Sustainability.

I. INTRODUCTION

The contemporary idea of nanomaterials was introduced by physicist Richard Feynman in his speech "There's Plenty of Room at the Bottom" in 1959, and it established new routes in the field of materials science. The early 21st century is characterized by an unprecedented shift in materials science, triggered by the development of nanomaterials. Nanomaterials are defined as materials that have at least one external dimension ranging from approximately 1 to 100 nm [1]. They exhibit novel and exciting electrical, mechanical, thermal, and optical properties that differ from their bulk counterparts. The size-dependent element of how their properties differ makes nanomaterials very advantageous for applications in broad fields. The advantages of nanomaterials over their bulk common counterparts fundamentally stem from properties related to their increased ratio of surface area-to-volume and quantum confinement effects, and this makes nanomaterials essential in fields that range from medicine and environmental remediation to electronics and batteries for energy storage [2].

Nanomaterials are generally synthesised in two ways, (a) top-down approach and (b) bottom-up approach. In the top-down approach, bulk precursors of large size are processed to generate nano-sized materials, while in the bottom-up approach, nano-sized materials are formed atom by atom or molecule by molecule [3]. The synthesis methods can also be classified into two major ways: (a) traditional synthesis methods and (b) green synthesis methods [4]. The traditional methods are usually particularly scalable and allow researchers a high level of control over the structure and morphology of materials typieally used in sensing, environmental remediation, energy, and biomedical applications. The major disadvantage of traditional methods is the extensive use of organic solvents that have detrimental effects on human health as well as the environment. As a result, researchers have turned to examine more health and environmentally suited methods of nanomaterial synthesis, which has led to green synthesis. The green synthesis of nanomaterials can be defined as a method of synthesis that uses a sustainable approach to lessen or eliminate the use of toxic chemicals, minimize the use of high-energy, and eliminate toxic by-products through the use of natural precursors.

This chapter will be an overview of various traditional and green synthesis methods for the preparation of nanomaterials, while explaining the pros and cons of each. Traditional methods, including physical or chemical methods, have many years of history in providing precise control of particle size and production of morphology. However, traditional methods commonly use toxic reagents and also higher energy inputs and produce hazardous by-products that can result in sustainability and risks to the environmental and human health. Green synthesis methods offer sustainable alternatives for the production of nanomaterials, which utilize biological resources (e.g. plant extracts, microorganisms, and natural polymers) that act as reducing agents and stabilising agents for nanomaterials. These methods are environmentally friendly and provide ease of use, cost efficiency, and increase the biocompatibility of the obtained nanomaterials. Overall, the chapter offers an understanding of the advantages or benefits and emerging progress of green synthesis as a suitable and sustainable alternative to traditional methods.

II. Traditional Synthesis Methods of Nanomaterials

There are two main processes which are used to synthesize nanomaterials, shown in Figure 1, which are top top-down and bottom-up approaches.

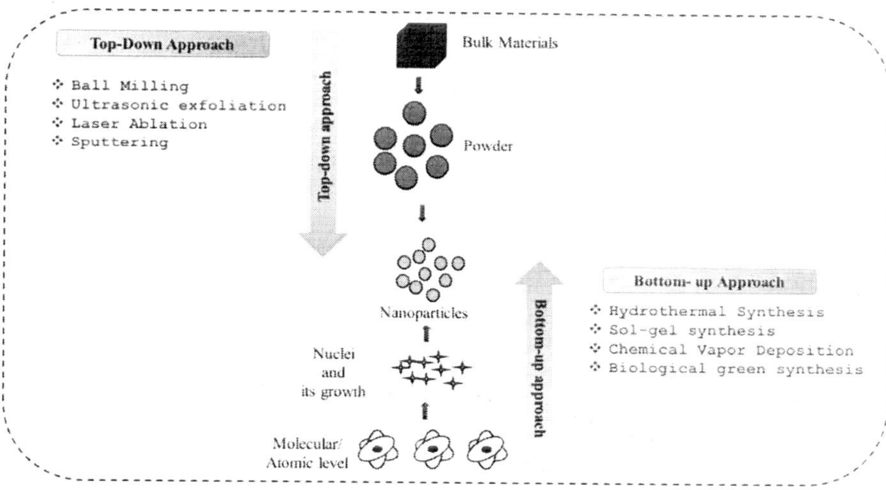

Figure 1: Nanomaterial Synthesis by Top-down and Bottom-up Approaches

A. Mechanical Ball Milling Technique

Around the end of the 1890s, Friedrich Fischer, a German physicist and chemist, was credited with the development of the ball mill [4]. Ball milling is a well-known method of synthesizing nanomaterials in powder form and is useful for the fabrication of nanostructured materials, composite systems, and the production of homogenous mixtures. Ball milling is a mechanical synthesis process and is widely used to reduce materials to the nanoscale and consequently increase their reactivity [5]. The ball milling process is accomplished using a ball mill, where the material and a stainless steel or ceramic grinding media are added into a rotating cylindrical chamber [6]. During the rotation of the chamber, which rotates at a consistent rpm, the ball media collide into the material of interest, resulting in shear and impact forces that minimize particle size and cause structural change. The planetary ball mill is the most common system to synthesize all types of materials, including composite and metastable materials.

The ball milling method is among the most environmentally friendly and sustainable methods for synthesizing nanomaterials. It is a greener alternative to many chemical routes because it requires no toxic solvents, hazardous reagents, or high-temperature conditions for nanomaterial production. However, it has some limitations when assessed against the general principles of green synthesis. The mechanical grinding involved in the ball milling process is very energy-intensive, and the duration of the process can outweigh some of the environmental benefits.

B. Ultrasonic-Assisted Exfoliation

The ultrasonic-assisted exfoliation or sonochemistry method is a much-debated technology employed in the synthesis of nanomaterials to produce nanomaterials and maintain uniform dispersion of nanoparticles within a solvent. This method involves the application of acoustic waves to induce a cavitation process that, in turn, generates chemical substrates required for the synthesis of nanoscale materials. During the cavitation process, physical interactions and chemical reactions take place between the bubbles and the precursors present in the medium, leading to the formation of the nanomaterials [7]. The morphology and size of the synthesized nanomaterial can be altered by optimizing parameters, including precursor type, reaction medium, and frequency or power of the ultrasonic waves applied [8]. The synthesis of nanomaterials using ultrasound or sonochemistry utilizes little, if any, organic solvents or harsh chemicals because the water may be used as a solvent, which is an environmentally friendly conventional synthesis procedure for nanomaterial synthesis. Additionally, the energy consumption required to create ultrasonic waves is much lower than that of several other synthesis techniques. However, a significant drawback of the ultrasonic method is the scalability; researchers continue to research to develop ultrasonic synthesis procedures at larger scales, generating the capability of working without the use of chemical catalysts.

C. Laser Ablation

The laser ablation technique allows for the fragmentation of bulk material at a high energy density from laser pulses into smaller sizes. The resultant nano-sized particles are formed in a gaseous or liquid medium that can regulate the size and shape of the synthesized nanomaterials. When synthesized via ablation in a gas or vacuum, the nanomaterials can deposit as nanomaterials on a surface to form a thin film [9] and, when ablated in a liquid medium, the resultant nanomaterials are stabilized into a colloidal suspension. In addition, the laser can be infinitely tuned through intensity, pulse duration, and wavelength to control its size and shape. The laser also has the unique ability to provide the ability of fine-tuned particle characterization and delivery control to the user, versus other traditional methods such as evaporation or ball milling [10, 11], and can also be done in a green manner, but it is frequently done in the presence of organic solvents to better control the morphology of the materials synthesized. Typical uses of laser ablation in organic solvents are usually for various metal nanoparticles [12]. This is one of the most energy-intensive methods to synthesize nanomaterials. The energy expenditure comes at a greater overall financial cost to the environment because this is often electricity that comes from coal- or gas-based sources to power the laser.

D. Hydrothermal Synthesis

Hydrothermal synthesis, also referred to as solvothermal synthesis, is the most commonly used synthesis method. Hydrothermal synthesis harnesses the solubility of the material by placing it under high temperature and pressure, resulting in the formation of a crystalline structure. In order to synthesise the desired products, the precursors, water (or another suitable solvent), and stabilising agents are mixed in a steel autoclave, which is subsequently heated and maintained over a fixed period of time. The morphology, size, and shape of the resulting nanomaterials can be tuned by changing parameters such as the precursor type, reaction temperature, or the pH of the solution. Following the run has completed, the autoclave is allowed to cool to room temperature and then washed and dried to achieve the final product. Hydrothermal synthesis, compared to many types of nanomaterial synthesis techniques, is generally more energy efficient and environmentally friendly because it occurs at relatively lower temperatures within the autoclave. Hydrothermal synthesis is a significant step forward for more sustainable and scalable nanomaterial synthesis; however, it does not fulfill all 12 Principles of Green Chemistry.

E. Sol-gel Method

The sol-gel method is a common method of synthesising nanomaterials. The method is relatively simple and economical and provides control over the composition and structure of the resulting nanomaterial. Generally, a sol refers to a liquid phase with a colloidal suspension of particles, and with hydrolysis and condensation (polymerisation), the system moves to the gel phase [15]. This process works especially well when producing metal oxide nanomaterials, yielding benefits such as a reduced particle-size distribution, high purity, and the applicability to coat complicated surface chemistries.

The method does have a few downsides concerning environmental and personal safety, in that many organic solvents that are conventional to hydrolyse nanomaterial precursors pose threats to personal health and the environment [16]. Overall, the sol-gel method is an effective and efficient method of producing nanomaterials, but it also presents elevated risks associated with its use.

F. Chemical Vapor Deposition

Chemical vapor deposition (CVD) is a common and effective method for producing high-quality 2D nanomaterials. In the CVD process, nickel, iron, and zinc act as a substrate and are subjected to one or more gaseous or vapor-phase precursors that react on their surface to yield a thin film or layered product [17, 18]. The reaction process typically occurs in the vacuum chamber at elevated temperatures (i.e., 1000°C or greater), often in the presence a nitrogen gas and a catalyst to aid in growth. The chemical vapor deposition process also allows researchers to manipulate a number of factors, such as the type of substrate, type and concentration of precursor, and temperature, in order to control the morphology, size, and general structure of the resultant material [19]. This nanomaterial synthesis process is rapidly being modified into a number of different variations of the original method, each having its own potential hazards in its use. The most common concern amongst these variations is the use of vapors and gases that can be very volatile. Often, the volatile gases and vapors themselves are not generally dangerous; however, the by-products of the vapors and gases when they contact the substrate or the catalyst can be potentially hazardous to both the environment and the user [20].

Other than the methods of synthesis discussed above, there are different synthetic methods of nanomaterials; for example, magnetron sputtering, molecular beam epitaxy, atomic layer deposition, thermal evaporation, electron beam evaporation, cathodic arc deposition (part of sputtering deposition), ion beam deposition (part of sputtering deposition) relevant to the physical vapor deposition technique, which typically uses thin film deposition.

III. Green Synthesis of Nanomaterials: Biological Approaches

This part focuses on biological green synthesis shown in Figure 2, which relies on living organisms, or their extracts, to reduce the size to form nanomaterials, or relies solely on the organism itself for nanomaterials production. These biological approaches utilize naturally occurring biomolecules (e.g., enzymes, proteins, flavonoids, polysaccharides) as reducing/capping agents in mild aqueous conditions (room temperature, neutral pH), which provide improved sustainability and safe environmental procedures [1]. In this section, we will focus specifically on green synthesis of nanomaterials with plant extract or microorganisms, including bacteria, fungi, algae, and enzymes, that serve as heterogeneous green resources.

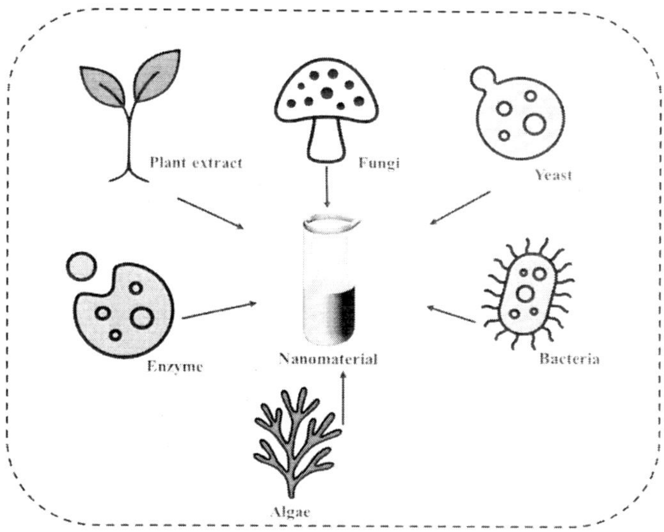

Figure 2: Different Green Biological Approaches to the Synthesis of Nanomaterials

A. Plant and its Extract-Based Synthesis

One of the most environmentally sustainable practices in green synthesis is to derive nanomaterials from plant sources or plant extracts. Typically, plants or plants extract comes from a process of drying and grinding the materials, and then boiling them in water for a certain duration. It can get filtered and then stored at 4°C [21]. In this scenario, extracts from different parts of the plants, such as leaves, roots, fruits, and agricultural residues as well (for instance, peels), are generally abundant in bioactive molecules such as polyphenols, flavonoids, alkaloids, and terpenoids that act like naturally occurring reducing agents and stabilize metal ions or metals during the formation of nanoparticles. The process typically involved drying and grinding of plant source or plant extract, followed by boiling in water to extract the phytochemical, filtering it, and later adding a metal precursor to the extract in relation to a metal based on the experimental design. Then, the mixture is incubated, typically at room temperature, during which time the biomolecules extract reduce the metal ions and subsequently stabilize newly formed metal nanoparticles. In addition to the benefits, the nanoparticles synthesized are sometimes heterogeneous, namely, a range of particle sizes can even occur. This is largely due to the compositional differences of the plant extracts, which can greatly change with the seasons of the plant, the species of the plant, or the growing conditions of the plant [22].

B. Bacteria-Based Synthesis

Bacteria are single-celled organisms with a cell wall but no true nucleus and complex organelles, and can provide a useful, green method for synthesizing nanomaterials. While some strains are pathogenic, many strains of bacteria are naturally occurring in the human body and are generally safe to work with under the bench environment. Strains like *Escherichia coli* (E. coli) and Bacillus subtilis are good candidates for green synthesis, since they are easy to grow and can be genetically manipulated. These two properties make them convenient and sustainable alternatives for making nanomaterials using green methods. When bacteria are utilized to synthesize nanomaterials, the bacteria are first grown aerobically until they reach a desired optical density. When the desired density is achieved, the bacterial cells and culture medium are mixed with a nanoparticle precursor. After the reaction is completed (usually after a color change, which is apparent), the cells are harvested by centrifugation at a high rpm for 20–30 minutes. The resulting supernatant contains a suspension of the synthesized nanomaterials [23]. The shape and dimensions of these nanoparticles are primarily dependent on two factors: the bacterial strain employed and the precursor material used. In bacterial systems, a range of biomolecules act as reducing and stabilizing agents during the conversion of metal ion-based precursors to nanoparticles. Biomolecules such as the amino acids tyrosine and tryptophan, found in cell wall proteins and the cytosol, are known to reduce metal ions and provide stability to nanoparticles. Similarly, sugars (aldoses and ketoses) also play a role as stabilizing agents in the reactions. These biomolecules create a protective capping layer around the nanoparticles, providing the characteristics needed for biocompatibility and non-toxicity to mammalian cells [24]. As a result of the natural interactions of biomolecules with metal ions, metal is reduced and self-assembled into stable and well-defined nanostructures such as spherical nanoparticles.

C. Fungi-based Synthesis

Fungi, including yeast, are eukaryotic organisms that feed by releasing digestive enzymes into their environment and absorbing the nutrients afterward. A characteristic of fungi is "chitin," a polymer derived from glucose, which reinforces their cell walls. These cell walls are largely composed of chitin with various enzymes and proteins, and all of these components serve an important role in making nanoparticles of various shapes, sizes, and compositions. The process of nanoparticle-making in fungi can be done both intracellularly and extracellularly. The typical process of nanoparticle-making in the case of fungi begins with growing and incubating fungal cultures in nutrient broth for about 72 hours to make biomass [25]. Afterward, the biomass is filtered and washed, and mixed with a nanoparticle precursor. Since fungal biomass is already growing in a nutrient-rich solution, after about 24 hours, single particles of nanoparticles will start to form in the solution. Similar to bacteria, the amino acids that are found in yeast cell walls serve as natural capping and stabilizing agents of the nanoparticles being synthesized. More importantly, studies have shown that the nanoparticles made by the fungal process are generally non-toxic for therapeutic purposes [26], which is in contrast to many nanoparticles made over the last few decades that are inherently toxic.

D. Yeast-based Synthesis

Yeasts are single-cell organisms of the fungal kingdom that, like bacteria, are quite simple. In fact, the most common type, Saccharomyces cerevisiae, is involved in fermentation, where it converts sugars into alcohol and carbon dioxide and is used in baking and brewing as well. However, not all yeasts are safe, as seen in Candida albicans, which causes serious infections in the blood-stream and other sites. To synthesize nanoparticles using yeast, when first providing yeast extracts, both commercial and homemade, with a precursor solution, were placed in a flask and allowed to incubate for 24 hours. After incubation, the solution was filtered through a sterile filter to separate the spores from the nanoparticles [27]. The proteins associated with the yeast cell wall, especially those that contain primary amine groups, were responsible for reducing sister agglomerated silver chloride into nanoparticles [28].

E. Algae-based Synthesis

Algae belong to the group of photosynthetic eukaryotic organisms and share several similarities with plants; however, they differ in both structure and function. Like plants, algae contain chlorophyll and can be unicellular or multicellular, growing primarily in aquatic environments. Algae differ from true plants in that they do not have stems, leaves, or vascular tissues. Algae can have varied effects on humans, from beneficial, like Spirulina (which has large amounts of nutrients), to harmful, like Anabaena, which can be toxic if ingested [29,30]. Over the past few years, many different species of algae have received attention for their unique ability to facilitate green synthesis of nanomaterials. To synthesize nanomaterials from various algae species, the algae are first dried completely and ground into a powder. This powder is combined with water and incubated for about 24 hours before it is filtered. The resulting filtered product contains bioactive compounds and is then mixed with the precursor for the nanomaterial and incubated at room temperature. Once the color visibly changes, the nanomaterial has been synthesized. The diversity of abnormal morphologies in synthesized nanomaterials is greater than that of bacteria and fungi.

F. Enzyme-based Synthesis

Enzyme-assisted biosynthesis is a biological process where microbes use specialized proteins, called enzymes, to synthesize nanomaterials. Biosynthesis is regarded as a green method of synthesis since, instead of traditional precursors with toxic chemicals, microorganisms produce stable nanomaterials based on natural precursors in conjunction with their enzymes. The process involves enzymes called reductases, reducing metals (Au^{3+} to Au^{0}, Ag^{+} to Ag^{0}) either intracellularly or extracellularly. In the extracellular reaction, the enzyme is excreted by the microbe and, thus, can form the nanoparticles outside the cell, while in the intracellular reaction, it occurs in either the microbe's cell walls or in the periplasm. The size and shape of the nanoparticles are influenced by factors like pH, temperature, and ion concentrations. Enzyme-assisted biosynthesis has the benefits of being inexpensive and biocompatible

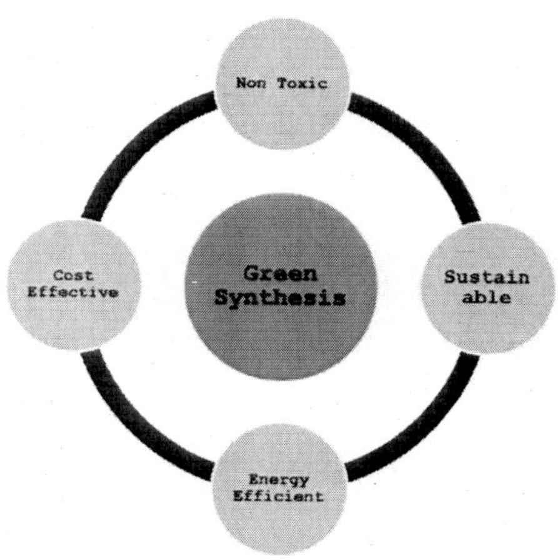

Figure 3: Advantages of Green Synthesis

and avoiding toxic chemicals, making it suitable for medical and environmental applications. This method also gives a certain level of control over nanoparticles' morphology, though the aforementioned benefits, the enzyme-assisted biosynthesis technique must overcome the obstacles of a sterile environment, desired uniform size distribution, and sufficient total yield quantity to generate a beneficial product on an industrial scale. This is often challenging, and these objectives may add complexity in terms of industrial utilization as well.

IV. Conclusion

The synthesis of nanomaterials offers a dynamic interplay between classical and green methods, with each method exhibiting various advantages and disadvantages. Classical methods (i.e., mechanical ball milling, ultrasonic-assisted exfoliation, laser ablation, hydrothermal and sol-gel methods, and chemical vapor deposition) result in better control of particle morphology and size and are better suited to applications in electronics, energy, and biomedicines; however, these classical methods use toxic reagents (the solvents) that are energy intensive and produce hazardous waste products.

Conversely, green synthesis methods – involving the use of plant extracts, bacteria, fungi, yeasts, algae, or enzymes combine to sustainably and conveniently generate biocompatible nanomaterials using biomolecules that can occur naturally within the sources. These sustainable methods use low amounts of toxic chemicals with very little energy to grow and produce the materials, and the methods are more economical; for these reasons, green synthesis is quickly becoming very attractive in medicine and environmentally related applications. There are, of course, large-scale challenges and inconsistency in the distribution of particle size, but the green synthesis methods remain perfectly positioned for innovative nanotechnology that is sustainable and responsible to the environment. Due to constant improvement in the optimization of these methods, green synthesis will be more established as a sustainable nanomaterial production method to fulfill the demands of globally eco-conscious initiatives.

Acknowledgement

R.L. would like to thank the DST Govt. of India for providing financial support under the INSPIRE scheme. H.S. would like to thank CSIR-HRDG for providing fellowship under the SRF-Direct Scheme. H.K. would like to acknowledge ISRO RAC-S, (GU/RAC-S/01/2020-21/03) for providing the financial support. H.K. would like to thank DST-PURSE (SR/PURSE/2022/116) for a research grant to Gauhati University.

REFERENCES

[1] B. Mekuye and B. Abera (2023). Nanomaterials: An overview of synthesis, classification, characterization, and applications, Nano select, 4(8), 486–501.

[2] A. Nazary, H. Shamloo and S. Mirzakuchaki (2025). Structural and Electronic Transport Properties of Functionalized Nanomaterials, Functionalized Nanomaterials for Electronic and Optoelectronic Devices: Design, Fabrications and Applications, 101–131.

[3] P. Nikolaidis (2020). Analysis of green methods to synthesize nanomaterials, Green synthesis of nanomaterials for bioenergy applications, 125–144.

[4] M. Huston, M. DeBella, M. DiBella and A. Gupta (2021). Green synthesis of nanomaterials, Nanomaterials, 11(8), 2130.

[5] K. Kubota and H. Ito (2020). Mechanochemical cross-coupling reactions, Trends in chemistry, 2(12), 1066–1081.

[6] S. Singh, H. Singh, R. Kumar, S. Dhiman, V. Kumar, R.S. Joshi and S.S. Gill (2021). A framework for effective and clean conversion of machining waste into metal powder feedstock for additive manufacturing.

[7] H. Xu, B.W. Zeiger and K.S. Suslick (2013). Sonochemical synthesis of nanomaterials, Chemical Society Reviews, 42(7), 2555–2567.

[8] J.H. Bang and K.S. Suslick (2010). Applications of ultrasound to the synthesis of nanostructured materials, Advanced materials, 22(10), 1039–1059.

[9] S.V. Zabotnov, A.V. Skobelkina, E.A. Sergeeva, D.A. Kurakina, A.V. Khilov, F.V. Kashaev, ... and M.Y. Kirillin (2020). Nanoparticles produced via laser ablation of porous silicon and silicon nanowires for optical bioimaging, Sensors, 20(17), 4874.

[10] Zeng, H., Du, X.W., Singh, S.C., Kulinich, S.A., Yang, S., He, J. and Cai, W. (2012). Nanomaterials via laser ablation/irradiation in liquid: a review, Advanced Functional Materials, 22(7), 1333–1353.

[11] V. Amendola and M. Meneghetti (2013). What controls the composition and the structure of nanomaterials generated by laser ablation in liquid solution? Physical Chemistry Chemical Physics, 15(9), 3027–3046.

[12] V. Amendola, P. Riello and M. Meneghetti (2011). Magnetic nanoparticles of iron carbide, iron oxide, iron@ iron oxide, and metal iron synthesized by laser ablation in organic solvents, The journal of physical chemistry C, 115(12), 5140–5146.

[13] Majumder, Sutripto (2023). "Nanostructured thin films by hydrothermal method." Simple Chemical Methods for Thin Film Deposition: Synthesis and Applications, Singapore: Springer Nature Singapore, 305–345.

[14] D. Nunes, A. Pimentel, R. Branquinho, E. Fortunato and R. Martins (2021). Metal oxide–based photocatalytic paper: A green alternative for environmental remediation, Catalysts, 11(4), 504.

[15] M. Parashar, V.K. Shukla and R. Singh (2020). Metal oxides nanoparticles via sol–gel method: a review on synthesis, characterization and applications, Journal of Materials Science: Materials in Electronics, 31(5), 3729–374.

[16] M. Tobiszewski, J. Namieśnik and F. Pena-Pereira (2017). Environmental risk-based ranking of solvents using the combination of a multimedia model and multi-criteria decision analysis, Green Chemistry, 19(4), 1034–1042.

[17] P.R. Somani, S.P. Somani and M. Umeno (2006). Planer nano-graphenes from camphor by CVD, Chemical Physics Letters, 430(1–3), 56–59.

[18] S. Park, S. Lim and H. Choi (2006). Chemical vapor deposition of iron and iron oxide thin films from Fe (II) dihydride complexes, Chemistry of materials, 18(22), 5150–5152.

[19] Y.B. Pottathara, Y. Grohens, V. Kokol, N. Kalarikkal and S. Thomas (2019). Synthesis and processing of emerging two-dimensional nanomaterials, Nanomaterials synthesis, 1–25.

[20] K.M. Choi, T.H. Kim, K.S. Kim and S.G. Kim (2013). Case study: Hazard identification of powder generated from a chemical vapor deposition process in the semiconductor manufacturing industry, Journal of occupational and environmental hygiene, 10(1), D1–D5.

[21] Y.H. Yap, A.A. Azmi, N.K. Mohd, F.S.J. Yong, S.Y. Kan, M.Z.A. Thirmizir and P.W. Chia (2020). Green synthesis of silver nanoparticle using water extract of onion peel and application in the acetylation reaction, Arabian Journal for Science and Engineering, 45(6), 4797–4807.

[22] N.I. Hulkoti and T.C. Taranath (2014). Biosynthesis of nanoparticles using microbes—A review, Colloids and surfaces B: Biointerfaces, 121, 474–483.

[23] S. Gurunathan, K. Kalishwaralal, R. Vaidyanathan, D. Venkataraman, S.R.K. Pandian, J. Muniyandi, ... and S.H. Eom (2009). Biosynthesis, purification and characterization of silver nanoparticles using Escherichia coli, Colloids and Surfaces B: Biointerfaces, 74(1), 328–335.

[24] J. Markus, R. Mathiyalagan, Y.J. Kim, R. Abbai, P. Singh, S. Ahn, ... and D.C. Yang (2016). Intracellular synthesis of gold nanoparticles with antioxidant activity by probiotic Lactobacillus kimchicus DCY51T isolated from Korean kimchi, Enzyme and microbial Technology, 95, 85–93.

[25] P.K. Sukul and C. Kar (2023). Green conversion methods to prepare nanoparticle, Bioinspired and green synthesis of nanostructures: A sustainable approach, 115–139.

[26] A.U. Khan, N. Malik, M. Khan, M.H. Cho and M.M. Khan (2018). Fungi-assisted silver nanoparticle synthesis and their applications, Bioprocess and biosystems engineering, 41(1), 1–20.

[27] N.A. Gow and B. Yadav (2017). Microbe Profile: Candida albicans: a shape-changing, opportunistic pathogenic fungus of humans, Microbiology, 163(8), 1145–1147.

[28] X. Liu, J.L. Chen, W.Y. Yang, Y.C. Qian, J.Y. Pan, C.N. Zhu, ... and D.P. Zhang (2021). Biosynthesis of silver nanoparticles with antimicrobial and anticancer properties using two novel yeasts, Scientific Reports, 11(1), 15795.

[29] S. Gogna, J. Kaur, K. Sharma, R. Prasad, J. Singh, V. Bhadariya, ... and S. Jarial (2023). Spirulina—An edible cyanobacterium with potential therapeutic health benefits and toxicological consequences, Journal of the American Nutrition Association, 42(6), 559–572.

[30] A. Igwaran, A.J. Kayode, K.M. Moloantoa, Z.P. Khetsha and J.O. Unuofin (2024). Cyanobacteria harmful algae blooms: causes, impacts, and risk management, Water, Air, and Soil Pollution, 235(1), 71.

CHAPTER 2

Biodegradable Green Composite Materials: Properties to Applications

Jyotishmoy Borah

Department of Chemistry, Bineswar Brahma Engineering College, Kokrajhar, Assam, India

E-mail: borahjm@rediffmail.com

ABSTRACT: A composite material is a material, which is produced from more than one constituent material known matrix and reinforcing agent having unique physical, chemical and mechanical properties. A modern dynamic world cannot imagine its development without having the concept of advancement in composite materials. Naturally, composite materials are light in weight & low density, high creep resistance, high strength, stiffness and corrosion resistance etc. Composite materials with man-made thermoplastic and thermoset polymers, rubbers and fibres such as glass fibre, nylon, rayon etc. and metal or metallic fibres are used as reinforcing agent, which are non-biodegradable in nature. Also, these synthetic materials pose many drawbacks towards the environment because they produce toxic gases and serious soil, water and air pollution after improper disposal. However, composite materials, which are produced from biodegradable green matrices reinforced with materials such as biodegradable polymeric or non-polymeric materials have unique physical, biodegradable, chemical, mechanical properties and are eco-friendly in nature. These biodegradable composites have potential applications in diverse fields such as packaging, biomedical devices, agriculture, construction materials, textiles and in many more. These materials contribute to reducing many synthetic conventional materials and mitigating environmental pollution through biodegradation and composting.

Keywords: Composite Material, Biodegradable, Green Composite, Fibres, Polymers.

I. INTRODUCTION

Over the last forty years plastics, polymers, ceramics and composites are the dominant materials, which are found to be extensive applications from household materials to industrial products. The attractive physical, mechanical and chemical properties of composite materials replace the conventional materials and hence they are widely used in many application fields such as transportation and infrastructure, sports, aerospace, automotive industry, environmental engineering, electrical distribution, and at many more areas.

Composite materials, the name represents a class of unique man-made material by combining two or more components having different properties to create new useful materials. The better or improved material characteristic of composite materials as compared to individual components is due to the dispersion of a reinforcing phase within a matrix [1]. Among the different polymer composites, fibre-reinforced polymer (FRP) composites can be regarded as alternative materials to metallic, non-metallic, polymer or ceramic materials due to their unique properties [2]. Different types of fibres such as carbon, aramids, glass and natural fibers are used as reinforcing materials while polymers such as polyester, phenolic, epoxy or vinyl ester are used as matrices in FRP composites [3-5]. The advantages of using FRP composites over metallic materials or polymer materials are low production cost, better corrosion resistance, higher specific strength and stiffness, non-reactive to chemicals and weather, higher fatigue resistance, better and controlled electrical and thermal properties, easier to mold and many more [6]. Since the matrices as well as the reinforcing materials used for the preparation of FRP composites are polymeric in nature hence there is a serious challenge of reuse and recycling. Although, thermoplastics polymer-based FRP composites can be recycled by melting, but thermoset polymer-based FRP composites are difficult to recycle due to their infusible property as well as insoluble in most of the common solvents [7].

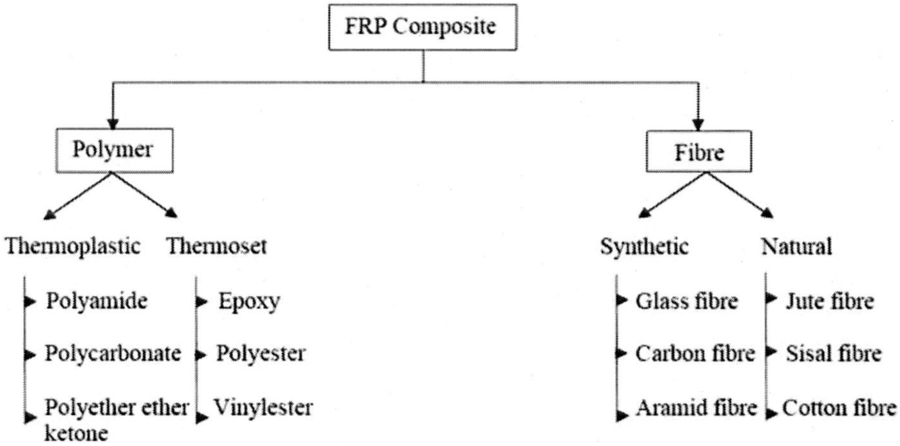

Figure 1: Commonly Used Materials for FRP Composites

However, in recent years, numbers of research articles have been published where bio-degradable polymers and natural fibres are used to develop composite materials instead of synthetic polymers and fibres [8, 9]. The primary advantage of using such raw materials to develop natural fibre reinforced composites (NFRCs) is the degradation after service life [10]. The NFRCs are also known as green composites, bio-composites or eco-composites. This type of composites is made of natural fibres as reinforcing materials embedded with the thermoplastics or in some cases thermoset plastics as matrices [11-13]. Natural fibres are readily available in nature and low-cost raw materials. Properties like bio-degradability, non-abrasivity, mechanically strong but less elastic, low density, high aspect ratio for efficient stress transfer and other high specific properties makes it more convenient than other conventional reinforcing materials [14, 15]. Based on origin, natural fibres are classified as plant, animal and mineral based fibres. A plant fibre mainly contains cellulose, which can be extracted from the leaf, husk, bast or stem, seed, hair of a plant. Cotton (seed hairs), ramie, jute, betel nut, flax (bast fibres), sisal and abaca (leaf fibres) are commonly used plant fibres [16, 17]. Out of these natural plant fibres, jute, flax, sisal, ramies are the most commonly used fibres for NFRCs. Animal fibres such as silk, hair, fur, wool etc. consist of largely certain proteins. However, recent studies show that the animal fibres have not much attracted as compared to plant fibres to develop NFRCs [18]. On the other hand, natural mineral fibres are non-metallic, inorganic fibres, which are directly derived from mineral sources mainly rocks. The most popular example of this type is asbestos. However, mineral based fibres particularly asbestos poses significant health hazards and hence very less interest have been observed as compared to plant and animal fibres.

From the literature reviews, it is observed that among the different class of natural fibres, the plant-based and few animal-based natural fibres have been widely utilized as reinforcing agent to fabricate the NFRCs.

In this article, some important properties and applications of specific plant and animal-based NFRCs are focused.

II. Reinforcing and Matrix Materials

NFRCs consists of mainly two constituent materials known as matrix, a polymer or plastic material and reinforcing agent called a natural fibre. Along with these two main constituent materials some secondary substances such as filler, fire retardant, colorant etc. are used to achieve the specific property for particular applications.

A. Reinforcing Materials

Since, synthetic fibres do not degrade easily so in the recent years, applications of various biodegradable fibres have been increasing. Most of the natural fibres used in FRP composite materials are plant-based, but some animal-based FRP composite materials have also been developed in the recent years [10]. The use of such natural fibres as reinforcing agent improves the sustainability of the NFRCs being constructed, extraordinarily in the automotive market.

1. **Plant-based fibres:** Plant based fibres are normally extracted from leaf, bast, seed, wood, fruits, grass and stalk. Among the natural sources, banana, pineapple, sisal, abaca (leaf fibre); jute, kenaf, ramie, hemp (bask fibre); cotton, linseed (seed fibre); bamboo, corn (grass fibre); coir, husk (fruit fibre) and rice, wheat (stalk fibre) are the most frequently used plant fibres [19, 20]. The three main components present in plant fibres are cellulose, hemicellulose, and lignine. High strength and stiffness of plant fibres is due to the crystalline structure of the cellulose. On the other hand, hemicellulose, which is highly branched polysaccharide easily interacts with cellulose thereby increasing the strength of the plant fibres. In addition to these, the phenolic compounds present in lignin binds with the cellulose retarding the microbial degradation of the fibres [21, 22].
2. **Animal-based fibres:** Hair, skin, fur or other parts of body are main sources of animal fibres. Broadly, animal fibres can be further classified as silk, wool and hair. The main chemical constituent of wool and hair is a protein called keratin, which is about 33% to 97% of total constituents. On the other hand, silk contains another protein called fibroin (75%) and sericin (22.5%). Except silk fibres, most of the animal fibres have low mechanical properties. Although silk has excellent mechanical properties but they are expensive. Due to these reasons animal fibres are not being used in FRC materials [23]. The significant properties of some commonly used plant and animal-based fibres are highlighted in the Table 1.

Table 1: Key Properties of Some Plant and Animal-Based Fibres [24-33]

Fibres	***Properties***	***Diameter (ʃm)***	***UTS* (MPa)***	***Elongation at Break (ʃm)***	***E** (GPa)***	***Density (g/cm³)***
Jute	Strength, durable	17–20	200–800	1.16–8	10–55	1.3–1.5
Flax	Lightweight, absorbent	12–16	300–1500	1–3–10	24.80	1.4–1.5
Sisal	Strength, durability	200–400	80–840	2–25	9–38	1.5
Pineapple	Soft, lightweight	10–28	170–1627	2.4	60–82	0.8–1.6
Banana	Warm, thick, durable	200	529–914	3	27–32	–
Ramie	Heavy, tough	20	384–938	1.2–8	44–128	–
Coir	Strength, durability	10–20	106–175	14.21–49	4–6	1.2
Kenaf	Rough	2.5–3.3	296–1191	3.5	2.86	–
Abaca	Thin, lightweight	40	980	–	7.30x10^{-4}	–
Hemp	Strength, durability	16–50	310–900	1.6–6	30–70	1.48
Wool	Warmth	16–40	120–174	25–35	2.3–3.4	–
Cotton	Lightweight, absorbent	11–22	264–800	3–8	5–12.6	–
Mulberry silkworm fibre	White-toned and more reproducible	9.5–10	208.45	19.6	6.10	1.33
Wild silk-worm fibre	Brownish toned	23–15	165.3	20.6	3.80	1.32
Twisted B. mori silk	–	10	248.8	33.5	5.8	–
Camel hair	Softness, warmth	20	212.15	37	3.9	–
Yak fibre	Warmth, softness, odor-resistant	15–19	270	14.53	45	3.4
Angora wool	Soft, thin fibers	12–16	–	–	–	
Alpaca	Lightweight, soft, fine, glossy, and luxurious	12–29	53.5	42.3	–	1.4
Catgut fibre		790	100	–	–	–

*Ultimate tensile strength, **Elastic modulus.

B. Matrix Materials

Based on properties, polymers can be broadly classified as elastomers, plastics and fibres. Due to low viscosity and high mechanical strength, plastics are utilized as matrix or continuous phase for FRC materials. Based on applications of FRC materials the matrix may be thermoset and thermoplastics. Glycidal-based epoxy such as diglycidyl ether of bisphenol-A (DGEBA), Novolac-based epoxy, vinyl ester resins, phenolic resins, polyesters are the commonly used thermoset plastics in FRC materials. The use of thermoplastics such as polyethylene, polypropylene, polycarbonates, polyamides, poly(vinyl chloride), polystyrene, poly(vinyl acetate), polylactic acid etc. have challenging issues as they do not have good mechanical and thermal properties as compared to thermoset plastics. In addition to these, they have high melting viscosity and processing temperatures. Since, they can be recycled after use, so the demand of these plastics as matrix components is being continuously increasing among the polymer scientists. The properties of some thermoplastics matrices are illustrated in Table 2 [34].

Table 2: Key Properties of Some Important Thermoplastics Polymers

Plastics	*Density ($g.cm^{-3}$)*	*Melting Temp. (°C)*	*Tensile Strength (MPa)*	*Tensile Modulus (MPa)*	*Thermal Conductivity $W.m^{-1}.K^{-1}$*
Low Density Polyethylene (LDPE)	0.91–0.92	105–115	12–15	110–450	0.33
High Density Polyethylene (HDPE)	0.94–0.97	130–140	28–41	1000	0.4–0.54
Polypropylene (PP)	0.89–0.92	160–170	30–50	1300–1800	0.11–0.28
Polycarbonates (PC)	1.17–1.45	–	55–75	2100–2400	0.19–0.20
Polyvinyl chloride (PVC)	1.3–1.45	160	34–62	2800	0.17
Polystyrene (PS)	1.04–1.06	240–260	30	1900–2900	0.033–0.04
Polyvinyl acetate (PVAc)	1–18–1.19	–	7–18	708	0.16
Polylactic acid (PLA)	1.24–1.3	150–180	21–60	3400–4200	0.13

Currently, green bio-composites containing natural fibres as reinforcing agents and biodegradable polymers such as poly(vinyl acetate), poly(vinyl alcohol), or polylactic acid as matrix materials are receiving increasing attention not only because they are cost-effective materials but also due to environmental concerns.

III. Properties of NFRCs

The properties such as physical, mechanical, electrical properties, bio-degradability, fire resistivity, melting temperature, glass transition temperature (Tg) etc. of NFRCs are depend not only on the type of matrix materials but also the nature of natural fibres (continuous or discontinuous form) used. The melting temperature and Tg of a composite depend mainly on the type polymer. On the other hand, different physical properties like colour, physical appearance, hardness, density, etc. of composites are mainly dominated by the matrix and other additives, fillers and inorganic or organic dyes, as well as fabrication process. However, properties like mechanical, biodegradability, etc. depend on fibres, polymers and additives.

A. Mechanical Property

The most important and widely studied property of FRCs is known as mechanical properties. This property determines the field of applications of any composite material. The variation of different mechanical properties of FRCs is determined by the nature of the components (matrix, fibre and filler) and the volume fraction of polymer matrix and fibre. Properties like tensile strength (TS), stiffness etc. are mainly determined by fibres but share strength is mostly influenced by the polymer matrix. It has been observed that the different mechanical properties of FRCs are significantly improved by using continuous fibres as well as the by increasing volume fraction of the fibres, which can be observed from two mathematical equations 1 and 2 respectively [35].

$$\int_{FRC} = \int_{f} V_{f} + \int_{m} V_{m} \qquad \ldots (1)$$

and

$$\Sigma_{FRC} = \Sigma_{f} V_{f} + \Sigma_{m} V_{m} \qquad \ldots (2)$$

Where, $\int$ and Σ are the tensile strength (TS) and elastic modulus (EM) of a composite. $\int_{f}$, Σ_{f}, V_f and $\int_{fm}$, Σ_{m}, V_m are the corresponding TS, EM and volume fractions of the fibre and the polymer matrix respectively.

Fibres orientation [36, 37] and fibres to polymer matrices interaction significantly control the mechanical properties of NFRCs. Since the interfacial adhesion between the hydrophilic cellulosic fibres and hydrophobic polymer matrices is very low, resulting in poor mechanical properties of NFRCs [38]. To enhance the interfacial adhesion several investigations have been carried out, where surface of fibres is modified chemically by means of acid, alkali, saline treatments with different natural fibres before use [39-51].

B. Thermal Property

Thermal stability of NFRCs is another important property primarily for processing and applications in specific field. Normally, cellulosic fibres are thermally not very stable and degrade relatively at lower temperature than the synthetic fibres. The thermal degradation of cellulosic fibres initially starts between 250–300°C. Natural fibres have great tendency to absorb moisture due to the presence of hydroxyl group in cellulose. This results in hydrolytic degradation and hence deteriorates mechanical properties [52]. Exposures of UV light leading to oxidation of cellulose and hence breakdown the cellulose fibres [53]. Microbial attack on cellulosic fibres is another serious concern, leading to a loss of structural uniformity and mechanical strength [54]. During manufacturing of NFRCs, higher temperature is required resulting degradation of cellulosic fibres. This degradation of natural fibres can be avoided by either chemical treatments or by hybridization with other materials. Dang *et al.* used ramie fibres coated with graphene oxide (GO) and observed that the crystallization temperature of GO-fibres modified FR/PP composite is higher than the unmodified FR/PP composite [55]. Similar work has also been carried out by Gouda *et al*, where bamboo fibres is modified with graphene naonoplatelets (GNP) to investigate the thermo-chemical properties of bamboo/epoxy composite [56].

C. Biodegradability of NFRCs

One of the major challenges of commercial materials is the durability, degradability and recycling. As discussed above NFRCs contains cellulosic fibres and hydrolysable materials (biopolymer with ester linkages, PLA, PVA etc.) are degraded under aerobic and anaerobic conditions. Normally, biodegradability of composite materials can be investigated by UV-radiation, soil burial and anaerobic digestion. Kim *et al.* investigated the biodegradability of poly(butylene succinate) bio-composites with rice-husk flour as reinforcing agent and observed that the percentage of weight loss during compost soil burial test was significantly higher than the natural soil burial test. Also, the degradability of the bio-composites was increased with increasing the volume content of bio-flour, because the bio-flour is easily attacked by microorganisms [57]. Liu *et al.* also studied the biodegradation pattern of jute fibre-based poly(butylene succinate) composites and observed that the weight loss of composites is higher than the pure jute fibres. They also investigated the effect of biodegradation on surface modification of fibres and also by changing diameter of fibres [58]. Gomez *et al.* investigated the relative biodegradability of conventional polymer-based composites with natural fibres composite materials by composting, soil incubation and anaerobic digestion. They observed that bio-based composites degraded to an appreciable extend while polymer-based composites did not show appreciable biodegradability [59]. Scaffaro *et al.* developed PLA based bio-composites with *Posidonia Oceanica,* a lignocellulosic filler and observed that the biodegradability of developed bio-composites is much faster than polymer [60]. Maio *et al.* developed PLA based bio-composites with three different lignocellulosic fillers and investigated the photochemical degradation of the composites [61]. Chee *et al.* also investigated the biodegradation of bamboo/kenaf fibres reinforced epoxy hybrid composites by UV-irradiation at elevated temperature and soil burial. They observed that during soil burial, due to the enzymatic action of microorganisms as well as absorption of moisture degradation occur leading to weight loss of the composites [62].

IV. Applications of NFRCs

Low cost, easily available, biodegradability, corrosion free, Light weight, flexible but mechanically strong plant and animal-based fibres are the major reasons to replace synthetic fibres in FRCs. At present, the applications of NFRCs are not restricted to automobiles, packaging, buildings and other constructions materials but they are extensively used in geo-textiles, medicine, agriculture, biochemical devices, sports, electronics, biosensors and in many more applications.

V. Summary and Future Prospects

In summary, some easily available plant and animal based biodegradable fibres as reinforcing agents to develop NFRCs have been briefly discussed. Although thermoset polymers such as epoxy resins, vinyl ester, polyester etc. are non-recyclable but due to their excellent fibre-polymer interaction properties they have been widely used as matrix materials. However, in recent years, some thermoplastics polymers, synthetic biodegradable polymers such as poly(butylene succinate) and limited numbers of biodegradable polymers such as PVA, PVAc etc. are used as matrix to develop biodegradable or green composites. It is also observed that although extensive researches have been carried out to develop plant-based bio-composite materials as well as their applications in different fields but there are very limited researches performed on animal-based bio-composite materials.

The major challenges to use both plant and animal fibres as reinforcing agent to develop NFRCs is its moisture sensitivity and high flammability, sensitive to UV-ray of sunlight, which deteriorates the material properties. Also, it is difficult to develop 100% bio-composite materials because the reinforcing agents are extracted from natural sources. Thus, extensive researches need to be required to develop fully biodegradable composite materials to fulfill the demand in commercial applications.

References

[1] F.C. Campbell (2010). Structural composite materials, Chapter 1: Introduction to composite materials.

[2] T. Gurunathan, S. Mohanty and S.K. Nayak (2015). A review of the recent developments in biocomposites based on natural fibres and their application perspectives. Compos: Part A, 77, 1–25.

[3] D.K. Rajak, P.H. Wagh and E. Linul (2021). Manufacturing technologies of carbon/glass fiber-reinforced polymer composites and their properties: a review, *Polymers*, 13, 3721.

[4] A.T. DiBenedetto (2001). Tailoring of interfaces in glass fiber reinforced polymer composites: A review, *Mater Sci Eng A*, 302, 74–82.

[5] T.P. Sathishkumar, S. Satheeshkumar and J. Naveen (2014). Glass fiber-reinforced polymer composites-a review, *J Reinf Plast Compos,* 33, 1258–75.

[6] M.A. Karim, M.Z. Abdullah and T. Ahmed (2021). An overview: the processing methods of fiber-reinforced polymers (FRPs), *Int J Mech Eng Technol,* 12, 10–24.

[7] G. Oliveux, L. Dandy O. and G.A. Leeke (2015). Current status of recycling of fibre reinforced polymers: review of technologies, reuse and resulting properties, Prog Mater Sci, 72, 61–99.

[8] D.N. Saheb and J.P. Jog (1999). Natural fiber polymer composites: A review, *Adv Polym Tech,* 18, 351–63.

[9] D. Chandramohan and J. Bharanichandar (2013). Natural fiber reinforced polymer composites for automobile accessories, *Am J Environ Sci.,* 9, 494–504.

[10] B. De, M. Bera, D. Bhattacharjee, B.C. Ray and S. Mukherjee (2024). A comprehensive review on fiber-reinforced polymer composites: Raw materials to applications, recycling, and waste management, Prog Mater Sci, 146, 101326.
[11] L. Mohammed, M.M.N. Ansari, G. Pua, M. Jawaid and M.S. Islam (2015), A review on natural fibre reinforced polymer composite and its applications, Int J Polym Sci, 2015, 243947.
[12] S.K. Ramamoorthy, M. Skrifvars and A. Persson (2015). A review of natural fibres used in biocomposites: Plant, animal and regenerated cellulose fibres, Polym Rev, 55, 107–162.
[13] T. Vaisanen, O. Das and L. Tomppo (2017). A review on new bio-based constituents for natural fibre-polymer Composites, *J Clean Prod*, 149, 582–596.
[14] D.L. Uma, S.S. Bhagawan and S. Thomas (1997). Mechanical properties of pineapple leaf fiber-reinforced polyester composites, *J Appl polym Sci.*, 64, 1739–1748.
[15] W.H. Zhu, B.C. Tobais and R.S.P. Coutts (1995). Banana fibre strands reinforced polyester composites, *J Mater Sci letters,* 17, 508–510.
[16] A.R. Sanadi, D.F. Caulfield, R.E. Jacobson and R.M. Rowell (1995), Renewable Agricultural Fibers as Reinforcing Fillers in Plastics: Mechanical Properties of Kenaf Fiber-Polypropylene Composites, Ind Eng chem Res, 35, 1889–1896.
[17] P.R. Hornsby, E. Hinrichsen and K. Tarverdi (1997). Preparation and properties of polypropylene composites reinforced with wheat and flax straw fibres: Part I Fibre characterization, *J. Mater Sci.*, 32, 1009–1015.
[18] E. Szczepanik, P. Szatkowski, E. Molik and K. Pielichowska (2024). The Effect of Natural Plant and animal fibres on PLA composites degradation process, *Appl Sci.*, 14, 5600.
[19] Q.T. Shubhra, A.K.M.M. Alam, M.A. Gafur, S.M. Shamsuddin, M.A. Khan, M. Saha, *et al.* (2010). Characterization of plant and animal based natural fibers reinforced polypropylene composites and their comparative study, *Fibers Polym*, 11, 725–31.
[20] S. Anidha, N. Latha and M. Muthukkumar (2019). Reinforcement of Aramid fiber with bagasse epoxy bio-degradable composite: investigations on mechanical properties and surface morphology, *J. Mater Res Technol*, 8, 3198–212.
[21] S.K. Ramamoorthy, M. Skrifvars and A. Persson (2015). A review of natural fibers used in biocomposites: Plant, animal and regenerated cellulose fibers, *Polym Rev.*, 55, 107–62.
[22] H.V. Scheller and P. Ulvskov (2010). *Hemicelluloses: Annu Rev Plant Biol*, 61, 263–89.
[23] M.M.R. Khan, A. Islam, M.A. Habib and M.N. Pervez, Raw silk processing to textile fibre and comparative analysis between acid and reactive dyed samples of silk fabric by laboratory experiment, *Int J Sci Eng Res*, 6, 19–27 (2015), ISSN 2229-5518.

[24] P. Joseph, V. Joseph and S. Thomas (1999). Effect of processing variables on the mechanical properties of sisal-fiber-reinforced polypropylene composites, *Compos Sci. Technol,* 59, 1625–1640.
[25] D.N. Saheb and J.P. Jog (1999). Natural fiber polymer composites: A review. Advances in Polymer Technology, *J. Polym Process Inst,* 18, 351–363.
[26] C.A.S. Hill and A. Khalil (2000). Effect of fiber treatments on mechanical properties of coir or oil palm fiber reinforced polyester composites, *J Appl Polym Sci,* 78, 1685–1697.
[27] M.S. Sreekala, M.G. Kumaran and S. Thomas (1997). Oil palm fibers: Morphology, chemical composition, surface modification and mechanical properties, *J Appl Polym Sci.*, 66, 821–835.
[28] S. Katori and T. Kimura (2002). Injection moulding of silk fiber reinforced biodegradable composites., *WIT Trans Built Environ,* 59, 1–9, ISBN 1-85312-904-6.
[29] C. Baillie, R. Jayasinghe (2004). Green composites: Polymer composites and the environment; Elsevier: Amsterdam, *The Netherlands,* 75–106, ISBN: 1845690397, 9781845690397.
[30] H.-Y. Cheung, K.-T. Lau, M.-P. Ho and A. Mosallam (2009). Study on the mechanical properties of different silkworm silk fibers, *J Compos Mater,* 43, 2521–2531.
[31] G. Singh, R.S. Dondapati and L.P. Singh (2021). Biomaterials printing for sustainability. In sustainability for 3D printing, *Springer: Berlin/ Heidelberg, Germany,* 15–28.
[32] A.R. Rant, O. Pofelska and W. Rant (2018), Characteristics of alpaca wool from farmed animals located on different continents, *Annals of Warsaw University of Life Sciences-SGGW-Animal Science,* 57,151–158.
[33] M.R. Sanjay and B. Yegesha (2017). Studies on natural/glass fiber reinforced polymer hybrid composites: An evolution, *Mater Today Proced,* 4, 2739–2747,
[34] M. Balasubramanian (2016). Introduction to composite materials. In fibrous and textile materials for composite applications, Springer: Berlin/ Heidelberg, Germany, 1–38, e-Electronic ISSN: 2197-9871.
[35] Y.J. You, J.H.J. Kim, S.J. Kim and Y.H. Park (2015). Methods to enhance the guaranteed tensile strength of GFRP rebar to 900 MPa with general fiber volume fraction, Constr Build Mater, 75, 54–62.
[36] J. Gassan, (2002). A study of fibre and interface parameters affecting the fatigue behaviour of natural fibre composites, *Compos Part A: Appl Sci Manuf,* 33, 369–374.
[37] M.R. Shetty, K.R.V. Kumar, S. Sudhir, P. Raghu, A.D. Madhuranath and R.M.V.G.K. Rao (2000). Effect of fibre orientation on mode-I interlaminar fracture toughness of glass epoxy composites, *J Reinf Plast Compos,* 19, 606–620.
[38] L. Yan, N. Chouw and X. Yuan (2012). Improving the mechanical properties of natural fibre fabric reinforced epoxy composites by alkali treatment, *J Reinf Plast Compos*, 31, 425–437.

[39] L. Boopathi, P.S. Sampath and K. Mylsamy (2012). Investigation of physical, chemical and mechanical properties of raw and alkali treated Borassus fruit fiber, *Compos Part B: Eng,*, 43, 3044–3052.

[40] O.M.L. Asumani, R.G. Reid and R. Paskaramoorthy (2012). The effects of alkali-silane treatment on the tensile and flexural properties of short fibre non-woven kenaf reinforced polypropylene composites, *Compos Part A: Appl Sci Manuf*, 43, 1431–1440.

[41] B.K. Goriparthi, K. Suman and N.M. Rao (2012). Effect of fiber surface treatments on mechanical and abrasive wear performance of polylactide/jute composites, *Compos Part A: Appl Sci Manuf*, 43, 1800–1808.

[42] E. Bisanda and M. Ansell (1991). The effect of silane treatment on the mechanical and physical properties of sisal-epoxy composites, *Compos Sci. Technol*, 41, 165–178.

[43] Y. Liu, X. Lv, J. Bao, J. Xie, X. Tang, J. Che, Y.H. Ma and J. Tong (2019). Characterization of silane treated and untreated natural cellulosic fibre from corn stalk waste as potential reinforcement in polymer composites, *Carbohydr Polym,* 218, 179–187.

[44] L.G. Dantas, L.A.D.C. Motta, D. Pasquini and J.G. Vieira (2019). Surface esterification of sisal fibres for use as reinforcement in cementitious matrix, *Mater Res,* 22, e20180585.

[45] M.A. Abu Bakar, S. Ahmad, W. Kuntjoro and S. Kasolang (2013). Effect of carbon fibre ratio to the impact properties of hybrid kenaf/carbon fibre reinforced epoxy composite, *Appl Mech Mater*, 393, 136–139.

[46] M. Aslan, M. Tufan and T. Kucukomerooglu (2018). Tribological and mechanical performance of sisal-filled waste carbon and glass fibre hybrid composites, Compos Part B Eng, 140, 241–249.

[47] P.N. Khanam, H.A. Khalil, M. Jawaid, G.R. Reddy, C.S. Narayana and S.V. Naidu (2010). Sisal/carbon fibre reinforced hybrid composites: Tensile, flexural and chemical resistance properties, *J Polym Environ,* 18, 727–733.

[48] G. Goud and R.N. Rao (2012). Mechanical and electrical performance of roystonea regia/glass fibre reinforced epoxy hybrid composites, *Bull Mater Sci,* 35, 595–599.

[49] K.R. Bharat, S. Abhishek and K. Palanikumar (2017). Mechanical property analysis on sandwich structured hybrid composite made from natural fibre, glass fibre and ceramic fibre wool reinforced with epoxy resin, *IOP Conf Ser Mater Sci Eng*, 205, 12015.

[50] A. Atiqah, M. Jawaid, M. Jawaid, M. Ishak, M. Ansari and R. Ilyas (2019). Physical and thermal properties of treated sugar palm/glass fibre reinforced thermoplastic polyurethane hybrid composites, *J Mater Res Technol,* 8, 3726–3732.

[51] S. Misri, Z. Leman, S.M. Sapuan and M.R. Ishak (2010). Mechanical properties and fabrication of small boat using woven glass/sugar palm fibres reinforced unsaturated polyester hybrid composite, *IOP Conf Ser Mater Sci Eng,* 11, 012015.

[52] R.N. Ibbett, D. Domvoglou and D. Phillips (2008). The hydrolysis and recrystallisation of lyocell and comparative cellulosic fibres in solutions of mineral acid, *Cellul*, 15, 241–254.

[53] K. Semthaun, R. Maliyam, C. Kaewdang and R. Buntem (2018). The thermal durability of the SiO_2-modified papers and photodegradation study of coated curcumin dye, *Mater Today Proc*, 5, 11074–82.

[54] B. Siroka, A.P. Manian, M.F. Noisternig, U. Henniges, M. Kostic, A. Potthast, U.J. Griesser and T. Bechtold (2012). Wash-dry cycle induced changes in low-ordered parts of regenerated cellulosic fibers, *J Appl Polym Sci*, 126, E397–408.

[55] C.Y. Dang, X.J. Shen, H.J. Nie, S. Yang, J.X. Shen, X.H. Yang and S.Y. Fu (2019). Enhanced interlaminar shear strength of ramie fiber/polypropylene composites by optimal combination of graphene oxide size and content, *Compos Part B Eng*, 168, 488–495.

[56] K. Gouda, S. Bhowmik and B. Das (2020). Thermomechanical behavior of graphene nanoplatelets and bamboo micro filler incorporated epoxy hybrid composites, *Mater Res Express*, 7, 015328.

[57] H.-S. Kim, H.-J. Kim, J.-W. Lee and I.-G. Choi (2006). Biodegradability of bio-flour filled biodegradable poly(butylene succinate) bio-composites in natural and compost soil, *Polym Degrad Stab*, 91, 1117–1127.

[58] L. Liu, J. Yu, L. Cheng and X. Yang (2009). Biodegradability of poly (butylene succinate) (PBS) composite reinforced with jute fibre, *Polym Degrad Stab*, 94, 90–94.

[59] E.F. Gomez and F.C. Michel (2013). Biodegradability of conventional and bio-based plastics and natural fiber composites during composting, anaerobic digestion and long-term soil incubation, *Polym Degrad Stab*, 98, 2583–2591.

[60] R. Scaffaro, A. Maio and E.F. Gulino (2021). Hydrolytic degradation of PLA/Posidonia Oceanica green composites: a simple model based on starting morpho-chemical properties, *Compos Sci Technol*, 213, 108930.

[61] A. Maio, E.F. Gulino, M. Gammino, M.C. Citarrella and R. Scaffaro (2025). Photochemical degradation of PLA-based green composites containing waste biomass from *Posidonia oceanica*, *Chamaerops humilis* and *Ailanthus altissima*: A comparative study, *Polym Degrad Stab*, 234, 111204.

[62] S.S. Chee, M. Jawaid, M.T.H. Sultan, O.Y. Alothman and L.C. Abdullah (2019). Accelerated weathering and soil burial effects on colour, biodegradability and thermal properties of bamboo/kenaf/epoxy hybrid composites, *Polym Testing*, 79, 106054.

CHAPTER 3

Development of *Antheraea assamensis*-based Fibroin Silk Scaffolds to Sustainable Healthcare Materials

Raunak Raj Singh[1] **and Lima Hazarika**[2*]

[1]Faculty of Sciences (Biotechnology) Assam Down Town University, Guwahati, Assam, India

[2]Department of Zoology, Assam Don Bosco University, Guwahati, Assam, India

*E-mail: lima.hazarika@dbuniversity.ac.in

ABSTRACT: Silk fibroin, a natural protein polymer produced by silkworms, exhibits exceptional properties such as mechanical strength, biodegradability, and biocompatibility, making it an outstanding candidate for biomedical applications. Among global silk species, Antheraea assamensis (Muga silk), endemic to Assam and Northeast India, is renowned for its distinctive golden hue, superior tensile strength, and RGD motifs that enhance cell adhesion. Despite these advantages, its potential in biomedical research remains underexplored. This study investigates the extraction, purification, and biomedical potential of Antheraea assamensis fibroin, focusing on its suitability for developing sustainable healthcare materials. The fibroin, composed of antiparallel β-sheets rich in glycine and alanine, was isolated through degumming, dissolution, and dialysis processes, yielding high-purity protein solutions suitable for biomaterial fabrication. The material's mechanical tunability, combined with inherent bioactivity, supports diverse formats - films, hydrogels, nanofibers, and sponges- enabling applications in tissue regeneration, wound healing, and targeted drug delivery.

Hydrogels derived from Antheraea assamensis fibroin demonstrated potential for controlled therapeutic release in cancer treatment, gene therapy, and precision medicine. Comparisons with Bombyx mori (mulberry silkworm) fibroin highlight Antheraea assamensis fibroin's enhanced mechanical stability and bioactivity, particularly for scaffolds requiring robust cellular interaction. Literature from Biman Mandal and others further supports the versatility of silk fibroin in skin repair, bone regeneration, vascular grafting, and neural engineering. The integration of traditional Muga silk culture with advanced biomaterials science positions Antheraea assamensis fibroin as a region-specific, eco-conscious alternative to synthetic polymers. However, clinical translation will require standardized processing methods, comprehensive in vivo biocompatibility testing, and scalable manufacturing. Overall,

Antheraea assamensis fibroin offers a rare combination of strength, biodegradability, and biological functionality, representing a sustainable, high-performance biomaterial with significant potential to advance regenerative medicine and next-generation therapeutic platforms.

This chapter presents a literature review involving numerically published work combined with previous laboratory experiments. The chapter does not report new quantitative data, since it uses published studies to assess the bio distribution of Antheraea assamensis fibroin.

Keywords: *Antheraea assamensis*, Sericin, Fibroin, Silk Scaffolds.

I. INTRODUCTION

Silk, a natural fibrous protein produced by silkworms and spiders, has been used for centuries as surgical sutures due to its strength, healing properties, and resistance to infection. Modern biomedical research further recognizes its potential in wound healing and tissue engineering, including bone, cartilage, tendon, and ligament repair [1]. Previous research highlighted the exceptional potential of *Antheraea assamensis* silk in tissue engineering, with demonstrated success in vascular grafts and corneal endothelium culture [2]. The integration of hydrogels with *Antheraea assamensis* silk aims to create a novel platform for targeted therapeutic delivery and regenerative medicine, offering promising applications for surgical and medical use. Silk, a natural biomaterial with a long history in medical applications, has shown remarkable promise due to its unique combination of mechanical properties, biodegradability, and ability to support cellular growth [3]. Among the various types of silk, the Indian endemic *Antheraea assamensis* silk stands out due to its superior mechanical and biological properties, including the presence of RGD motifs that promote cell adhesion. Despite its potential, *Antheraea assamensis* silk remains underexplored in the field of tissue engineering, presenting an opportunity to harness its unique characteristics for advanced medical applications [4]. The *Antheraea assamensis* – based hydrogels, with high specificity and affinity for biological targets, introduce a novel dimension to *Antheraea assamensis* silk-based grafts in regenerative medicine. This paves the way to develop fibroin-based biodegradable materials extracted from *Antheraea assamensis* silk, which can be used in sustainable materials for tissue engineering [4,5].

Antheraea assamensis, commonly known as the Muga silk moth, is a large, wild silk-producing moth native to the Brahmaputra Valley and adjoining regions of Northeast India, particularly Assam. It belongs to the family *Saturniidae* and is renowned for producing the golden-yellow Muga silk, which is exclusive to this region and has been granted Geographical Indication (GI) status [4].

The adult moth has broad wings with characteristic eye-like spots and a wingspan that can exceed 10 cm. The larval stage feeds primarily on leaves of *Persea bombycina, Litsea monopetala*, and other *Lauraceae* plants. The golden lustre of Muga silk intensifies with age and washing, making it highly valued for traditional Assamese attire such as mekhela chadors and gamosas [5].

The life cycle of *Antheraea assamensis* includes egg, larva, pupa (cocoon), and adult stages, and it is reared in open semi-domesticated conditions. This species is sensitive to environmental changes, with temperature, humidity, and host plant availability influencing its growth and silk quality [5]. Due to its restricted distribution and dependence on specific host plants, *Antheraea assamensis* is considered a regionally significant but vulnerable silk moth. Conservation of its habitat and host plants is essential to sustain traditional Muga silk culture and associated rural livelihoods [6].

In this chapter, multiple sources of literature regarding *Antheraea assamensis* silk fibroin will be critically reviewed and synthesized, including how *Antheraea assamensis* silk fibroin is extracted, the resulting structure, and areas where *Antheraea assamensis* silk fibroin may be used in biomedicine [6-10]. The focus will not be to provide new experimental results, but rather to review published studies as well as relevant laboratory findings from the author's previous exploratory work and put these together to provide context on how materials and processing methods relate to potential uses of these materials.

II. Silk Fibroin and Extraction Methods of Silk Fibroin

Silk fibroin is an insoluble protein that is present in silk, and this fibroin is produced by silkworms. Amongst the silkworms, *Antheraea* and *Bombyx mori* proteins are widely used in molecular biology and bioengineering for the production of fibroin-based hydrogels, which are widely used in experiments. Fibroin extracted from silk is generally composed of three chains, which are glycoprotein P25, light (L) chain and heavy (H) chain, respectively. The heavy and light chains are connected through a disulfide bond, while P25 binds to this disulfide-linked pair via noncovalent interactions. P25 is crucial for preserving the structural integrity of the complex [11].

The heavy fibroin protein is composed of stacked layers of antiparallel β-sheets. Its primary structure is dominated by the repeating amino acid motif *(Gly-Ser-Gly-Ala-Gly-Ala)*$_n$. The abundance of glycine, and to a lesser extent alanine, enables dense sheet packing, which imparts silk with its rigidity and high tensile strength. This unique balance of stiffness and toughness makes it suitable for a wide range of applications, from biomedical uses to textile production [12].

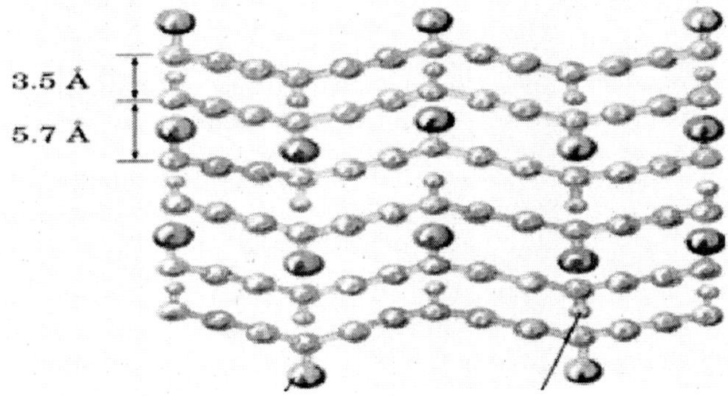

Figure 1: Structure of Fibroin

Source: https://www.researchgate.net/figure/Structure-of-fibroin_fig27_316139245

III. Overview of Silk Fibroin Extraction Methods (Based on Published Protocols)

The extraction steps described below summarize standard protocols reported in the literature, supplemented by representative images from earlier laboratory practice by the researcher [13].

Silk fibers consist of two major protein components Silk fibers consist of two major protein components: fibroin, which forms the structural core, and sericin, a gummy protein that surrounds and binds the fibroin fibers. To obtain fibroin, the removal of sericin is mandatory. [14] The process is also known as the 3D process: Degumming, Dissolution, Dialysis and this extraction protocol follows the following:

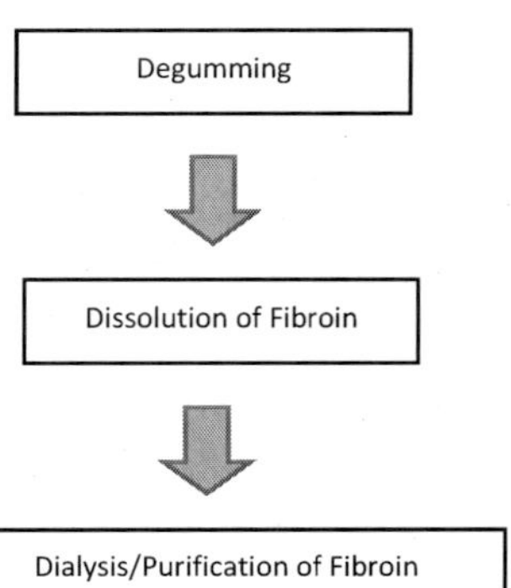

The Sericin removal or degumming is the most important step, as it separates the fibroin from the sericin coating, and the general common methods that are applied for this step are:

1. **Alkaline Soap Boiling:** Cocoons are boiled in a dilute soap or sodium carbonate solution to hydrolyse and wash away sericin.
2. **Enzymatic Degumming:** Proteolytic enzymes such as protease or papain are used to selectively digest sericin without damaging fibroin.
3. **High-Temperature Water Treatment:** Repeated boiling in distilled water under controlled conditions.

The dissolution of fibroin fibres is done right after the degumming, where the insoluble fibroin fibres are dissolved in strong solvents or salt solutions to break up the strong hydrogen bonds and the crystalline structures. The most common solvents that are usually involved are: Ajisawa's Reagent: A ternary mixture of calcium chloride, ethanol, and water [15].

Lithium Bromide (LiBr) Solution: Widely used for laboratory-scale fibroin dissolution.

Calcium Nitrate Hydrate or Cuprammonium Solutions: Used in early experiments, however, it is becoming less common due to toxicity concerns.

After the dissolution, Purification (Dialysis) is done where the fibroin solution is injected or placed in a dialysis membrane and immersed in distilled water for 3 days, and changing water every 6 hours to remove salts and residual chemicals. The result is an aqueous silk fibroin solution suitable for biomedical, textile, or material applications [16].

As the author has done some previous work in the extraction of fibroin for previous Experiments, it has been shown in the following images below:

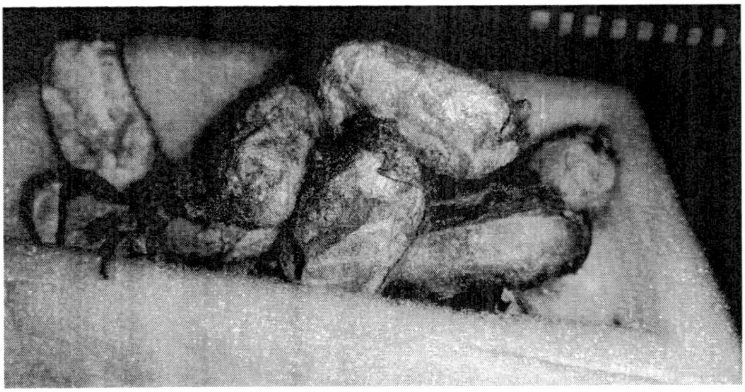

Figure 2: Antheraea assamensis Cocoons

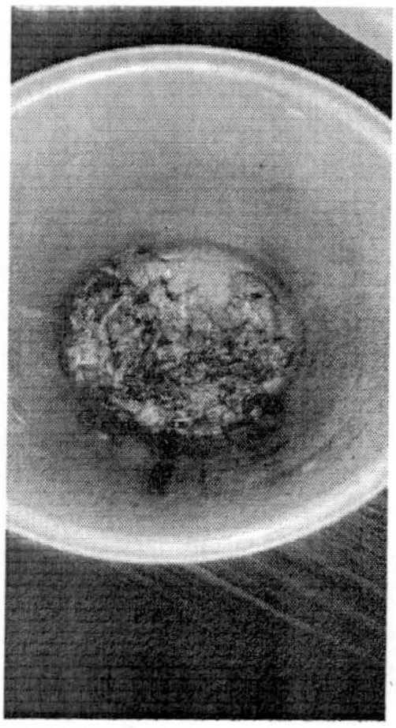

Figure 3: Degumming of *Antheraea assamensis* Cocoons

Figure 4: Dialysis of *Antheraea assamensis* Cocoons to Extract Fibroin

Figure 5: Fibroin Hydrogel Extracted after Dialysis of *Antheraea assamensis* Cocoons

IV. Utility of Silk Fibroin Extracted from *Antheraea assamensis* Hydrogel for Biomedical Applications

As per the previous findings of some eminent researchers who have worked in non-mulberry silks where *Antheraea assamensis* also falls in this category, the prospects of using *Antheraea assamensis* fibroin as a biomaterial is high for tissue engineering especially grafts (kin, cartilage, bone, vascular grafts). The key aspect of using non-mulberry silk fibroin like *Antheraea assamensis* can be resonated that the intrinsic cell binding motifs are better in *Antheraea assamensis* though further studies are required as the comparative studies are done less in some areas: for example the RGD sequences.

Earlier studies in the aspects of Skin/Wound healing, it has been observed that nanofibrous mats of the silk fibroin coated with recombinant spider-silk proteins having antimicrobial peptides and cell-binding/growth-factor peptides. These mats reduced bacterial population ~4-fold in vitro and helped produce vascularised bio-artificial skin grafts. The main focus was to treat chronic wounds like diabetic ulcers and burn victims skin grafts. The silk matrix is supportive of human dermal, epidermal and endothelial cells (for vascularisation) [16,17].

In terms of Cardiac/Heart Tissue Engineering, A study a conducted by Mandal *et al* showed that *Antheraea assamensis* had superior mechanical strength and low immunogenicity and had a better support system for the maturation for cardiomyocytes (e.g., up-regulation of *MYH6* (myosin heavy chain-α), *GJA1* (connexin 43)) [18].

It was also seen that *Antheraea assamensis* supported higher chondrocyte attachment, proliferation, sGAG production, Collagen II expression and most importantly they were biocompatible in vivo for cartilage tissue engineering, especially when compared with mulberry and other non-mulberry cocoons. A further study was also conducted later where the same set of researchers developed hierarchically structured seamless silk scaffolds for osteochondral interface (i.e., cartilage + bone) tissue engineering [18]. The scaffold used silk fibroin (both mulberry & non-mulberry), designed to have a biphasic architecture (fiber-free for cartilage, fiber reinforced for bone). The results in rabbits models gave a conclusion that the scaffold had enhanced cartilage + subchondral bone regeneration. These experiments show the potentialities of the *Antheraea assamensis* extracted fibroin for cartilage/osteochondral Engineering and the high efficiency rates which can be used for bone regeneration [18].

Recent studies on immunological tests also show that these scaffolds show different macrophage activation responses. *Antheraea assamensis* fibroin supported a macrophage response [18].

Other research on *Antheraea assamensis* fibroin also points out the nanoparticles made from this fibroin into Nano-delivery carriers when prepared from the extracted fibroin, which can be used for drug delivery [18]

V. Significance of *Antheraea assamensis* Fibroin Research

The non-mulberry silks like *Antheraea assamensis* are very much less utilized biomaterials on a global scale so many of these aspects aren't fully studied well in regards to inherent bio functionality like the RGD (cell binding motifs), which enhance cell adhesion without the requirement of synthetic peptides. The studies of *Antheraea assamensis* fibroin can be very well utilized in multi-tissue engineering like skin, cartilage, bone, and heart showcasing the versatility as a biomaterial in different tissues. Being endemic to Assam and NE India, this indigenous silk makes the work more feasible in terms of cost, availability, and ecological compatibility.

The studies and works summarized in this study indicate the aspects of *Antheraea assamensis* fibroin as a biomaterial from NE India for bioengineering and sustainable healthcare applications. While fibroins from other silks extracted from *Bombyx mori* and other silks have been extensively studied and the areas of biomedical research and bioengineering, *Antheraea assamensis* silk fibroin is still very much underexplored for the areas of mechanical and biological properties. The unique presence of RGD motifs in the structure of *Antheraea assamensis* fibroin, which is rare in other silk fibres, *Antheraea assamensis,* enhances cell adhesion, which is necessary for tissue engineering and regenerative medicine. The inherent bioactivity of *Antheraea assamensis* fibroin enables scaffolds to potentially require fewer surface modifications or coating processes, making the scaffold production easier.

From a sustainability perspective, *Antheraea assamensis* fibroin is an opportunity for regionally specific biomedical materials to be produced from Muga silk cultivars in Northeast India, further advancing global initiatives to produce biomaterials in an eco-friendly and sustainable manner. Sustainability initiatives help minimize dependence on externally produced synthetic polymers and reduce the carbon footprint of material production. However, maintaining sustainable sericulture practices and conserving the habitat of *Antheraea assamensis* remain vital, as the species is sensitive to environmental fluctuations and relies on specific host plants.

VI. Conclusion

Investigating *Antheraea assamensis* fibroin highlights its great potential as a high-performance, sustainable biomaterial for advanced healthcare. The unique features of *Antheraea assamensis* fibroin, such as its remarkable RGD motifs, mechanical properties, biocompatibility, and biodegradability, provide advantages over silks commonly used in laboratories, such as *Bombyx mori*. The versatility of *Antheraea assamensis* fibroin in being processed into different formats, including hydrogels, nanofibers, films, and sponges, also affords it the ability to meet a considerable range of bioengineering applications, such as tissue regeneration, wound healing, and localized delivery of therapeutics. By merging traditional Muga silk heritage with contemporary biomedical research, *Antheraea assamensis* fibroin offers a locally sourced, eco-friendly solution for regenerative medicine. However, successful clinical translation will depend on scalable extraction methods, long-term biocompatibility validation, and strong interdisciplinary collaboration

Acknowledgement

I would like to acknowledge my gratitude to Dr. Lima Hazarika, Assistant Professor, Department of Zoology, Assam Don Bosco University, for her persistent guidance, for this study. Dr. Lima Hazarika's guidance led the direction of this work.

Also, I acknowledge to Assam down town University for equipping the laboratory facilities that made this research possible.

REFERENCES

[1] T. Yamamoto, T. Miyajima, K. Mase and T. Iizuka (2003). Production of sericin a silk protein by the silkworm. *Bio Industry*, 20, 13–18.

[2] M. Gogoi, A. Gogoi and B. Baruah (2017). "Exotic muga silk: Pride of Assam" *International Journal of Applied Home Science*, Vol. 4, (1 & 2):72–78.

[3] S.C. Kundu, B.C. Dash, R. Dash and D.L. Kaplan (2008). Natural protective glue protein, sericin bioengineered by silkworms: Potential for biomedical and biotechnological applications. *Prog Polym Sci.* 33: 998–1012.

[4] F.G. Omenetto and D.L. Kaplan (2010). New opportunities for an ancient material. Science, 329(5991), 528–531.

[5] S.B. Dandin and S.N. Kumar (2007). Bio medical uses of silk and its derivatives. Indian Silk, 45(9), 5–8.

[6] R. Ahmad, A. Kamra and S.E. Hasnain (2004). Fibroin silk proteins from the non-mulberry silkworm Philosamia ricini are biochemically and immune-chemically distinct from those of the mulberry silkworm Bombyx mori. DNA Cell Biol.;23: 149–154.

[7] A.K. Gupta, K. Mita, K.P. Arunkumar and J. Nagaraju (2015). Molecular architecture of silk fibroin of Indian golden silkmoth, *Antheraea assama. Sci Rep.;* 5: 12706.

[8] A. Datta, A.K. Ghosh and S.C. Kundu (2001). Differential expression of the fibroin gene in developmental stages of silkworm, Antheraea mylitta (Saturni idae). *Comp Biochem Physiol B Biochem Mol Biol.;* 129: 197–204.

[9] S. Hazra, S. Nandi, D. Naskar, *et al.* (2016). Non-mulberry silk fibroin biomaterial for corneal regeneration. *Sci Rep.;* 6:21840.

[10] D. Gupta, A. Agrawal and A. Rang (2014). Extraction and characterization of silk sericin. *Indian J Fibre Text Res* 39: 364–372.

[11] M. Sasaki, N. Kato, H. Watanabe and H. Yamada (2000), Silk protein, sericin, suppresses colon carcinogenesis induced by 1, 2-dimethylhydrazine in mice. *Oncol Rep.,* 7: 1049–1052.

[12] I.B. Khalifaa, N. Ladharib and M. Touayc (2012). Application of sericin to modify textile supports. *J Text Inst* 103: 370–377.

[13] Y. Takasu, H. Yamada and K. Tsubouchi (2002). Extraction and chromatographic analysis of cocoon sericin of the silkworm, *Bombyx mori. J Insect Biotechnol Sericol.,* 71: 151–156.

[14] G.H. Altman, F. Diaz, C. Jakuba, T. Calabro, R.L. Horan, J. Chen ... and D.L. Kaplan (2003). Silk-based biomaterials. Biomaterials, 24(3), 401–416

[15] S. Sofia, M.B. McCarthy, G. Gronowicz and D.L. Kaplan (2001). Functionalized silk based biomaterials for bone formation. *Journal of Biomedical Materials Research: An Official Journal of The Society for Biomaterials and The Japanese Society for Biomaterials,* 54(1), 139 148.

[16] P. Aramwit, S. Kanokpanont, T. Nakpheng and T. Srichana (2010). The effect of sericin from various extraction methods on cell viability and collagen production. *International Journal of Molecular Sciences,* 11(5), 2200–2211.

[17] X. Hu, Q. Lu, D.L. Kaplan and P. Cebe (2009). Microphase separation controlled â-sheet crystallization kinetics in fibrous proteins. *Macromolecules,* 42(6), 2079–2087.

[18] B.B. Mandal and S. Kundu (2008). A novel method for disso lution and stabilization of non-mulberry silk gland protein fibroin using anionic surfactant sodium dodecyl sulfate. *Biotechnology Bioengineering,* 99: 1482 1489.

CHAPTER 4

Synthesis and Characterization of Copper Sulphide Nanomaterials by a Green Chemical Route in Polymer Matrices for Photovoltaic Device Applications

S.K. Nath

Department of Physics, B.H. College, Howly, Gauhati University, India

E-mail: sanatankrnath@gmail.com

ABSTRACT: A simple chemical route was adopted to synthesize copper sulphide nanoparticles using Cu (acc.) and thiourea as precursors with starch as a green capping material. The synthesized CuS exhibited polycrystalline hexagonal structure as observed from XRD analysis. Spindle-shapenanoparticle distribution was observed in large areas using HRTEM. The UV-Vis spectra showed absorption that shifted towards the blue region while the PL spectra show UV-blue emission with lowering molar concentration confirming quantum confinement. Optical band gap increased from 3.31 eV to 3.73 eV with lowering molarities as estimated from Tauc-plot. The particle size that was estimated from XRD result employing D-S and W-H plot agreed well with those of HRTEM results. Starch as capping materials played key roles in quantum confinement of nanoparticles. The FTIR showed a characteristic Cu-S stretching peak of copper sulphide at 615 cm^{-1}. The electrical properties of ionic copper sulphide solution suited well in determining the semiconducting nature of the synthesized material. The material synthesized may be useful for an optoelectronic device like UV-blue LED.

Keywords: Green Matrix, Spindle-Shape Nanoparticle, UV-Blue Emission, Ionic Conduction.

I. INTRODUCTION

In recent years nanocrystaline copper sulphide has been an important topic of research amongst the researchers working in the field of materials science. CuS, CdS, ZnS, HgS and PbS in bulk form behave as metal sulphides. But in nanoform these sulphides exhibit properties which are completely different from those of their bulk counterparts. Nanoscopic copper sulphide has been an extensive research work owing to its novel structural, optical and electrical properties [1-2]. Copper sulphide is a p-type semiconducting material with a narrow band gap and high absorption coefficient (10^4 cm^{-1}) [3-6]. It converts to superconductors at 1.6 K [7]. Another remarkable property of copper sulphide is its ability to form sub-stoichiometric compounds such as CuS, $Cu_{1.96}S$, $Cu_{1.8}S$, Cu_7S_4 and Cu_2S [8, 9]. Out of these, the amorphous brown chalcocite (Cu_2S) and the green-black crystalline covellite (CuS) phases are of special interest owing to their metal-like electrical properties [10]. Copper sulphide shows high emission and absorption in near IR region (800 nm–1500 nm) [11] and also exhibits fast ion conduction at high temperature [12]. Besides, copper sulphide exhibits size tunable properties depending on growth parameters such as molar concentration, temperature, capping materials, precursor solution composition, growth time, pH value of solution etc. The synthesized p-type layer of copper sulphide with a suitable n-type layer may be utilized to form a high efficiency semiconductor diode which may find its applications in various optoelectronic devices such as gas sensors, photovoltaic devices like solar cells, UV-blue LEDs and a few to mention.

From amongst a multitude of synthetic routes, a green chemical synthesis method has been employed to synthesize copper sulphide nanoparticles with different shape and size. The method is simple, cost effective, ecofriendly, nontoxic and requires low cost infrastructure. Copper acetate monohydrate and thiourea as precursor material for copper and sulphur source respectively and a green, nontoxic and eco-friendly capping material starch, a good bio-polymer with –OH group in its structures [13] have been used in the synthesis process. The –OH group of polysaccharides helps to obtain uniform structure as well as to protect the nanoparticles from aggregation. Starch plays a key role in confinement of nanoparticles as well as in protection of nanoparticles on the highly reactive surface of atoms for long period against ambient atmosphere. This type of polymer was also used by others. H.T. Boey *et al.* obtained colloidal chitosan-stabilized copper sulphides and observed that an increase in chitosan concentration resulted in smaller particle size [14]. M.A. Sangamesha *et al.* reported a facile aqueous synthesis of CuS nanoparticles of various sizes and shapes and found an increase in band gap of copper sulphide on decreasing molar concentration. J.Z. Zhang reported increase of band gap with decrease of size of nanoparticles while studying dynamic properties of charge carriers in nanoparticles [15]. A. Jamie *et al.* reported of obtaining the separate contributions of electron and ion transport to the electrical conductivity of thin film samples comprising carbon black mixed with a polymer electrolyte and concluded that ionic conductivity increased with increasing relative humidity [17]. F. Zhuge *et al.* while synthesizing copper sulphide thin films by successive ion layer absorption and reaction method found that larger band gap of copper sulphide films lead to decrease in ionic conductivity which is in fairly well agreement with our results [18]. The ionic conductivity study in an electrolytic ionic solution is found to be much suitable in determining the electronic semiconducting properties of copper sulphide nanoparticles.

The purpose of thepresent work was to prepare copper sulphide hexananoplate and nanoparticles in starch matrix by a green chemical method and to study the ionic conductivity of the product in an aqueous medium.The method is solution process able, scalable and stable over a long period against ambient atmosphere, easier to synthesize and characterize, environment friendly and inexpensive.

II. Experimental Details

Initially 3% starch solution was prepared dissolving in 100ml deionised water in a container using magnetic stirrer at room temperature. Then copper acetate monohydrate ($(CH_2COO)_2Cu.H_2O$) and thiourea (H_2NCSNH_2) of equal molar concentration (0.1 M) were dissolved in 3% starch mixed deionsed water in the volumetric ratio 1:2 in two separate containers under continuous stirring by a magnetic stirrer. Ammonia (aq.) as a complexing agent was added to the starch mixed copper ion (Cu^{++}) complex to make the solution alkaline until it becomes clear and homogeneous adjusting pH to 11 where thiourea solution was added drop wise for the formation of final copper sulphide matrix solution. The synthesized sample turned golden brown in color with black precipitates and was encoded as S1. Following similar procedures the other samples (S2-S10) were prepared. The synthesis scheme has been illustrated in Figure 1.

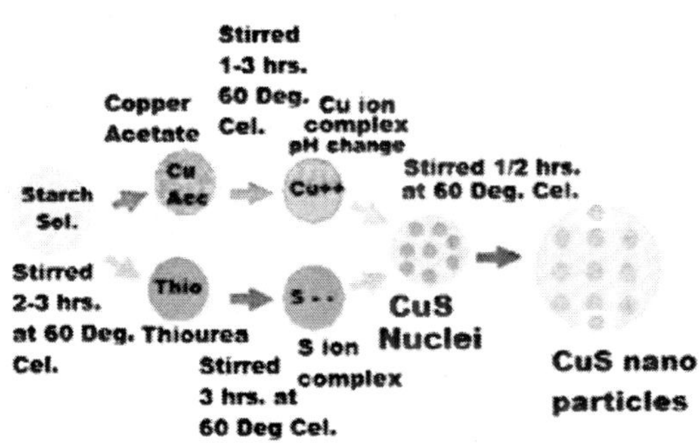

Figure 1: Schematic Synthesis Scheme of CuS Nanoparticles in Starch Matrix

XRD and high resolution electron microscopy (HRTEM) techniques were used to determine the structural characterization of the synthesized nanoparticles. The X-ray diffractograms were recorded using Cu Kα radiation (λ = 1.5406 Å). The X-ray tube was operated at 40kV and 40 mA. The 2θ was varied from 0° to 80° with a scanning rate 0.05°. The HRTEM morphology was recorded using JEM 2100 (Jeol Electronic Microscope) at accelerating voltage 200 kV and beam current 102 μA. Selected area electron diffraction (SAED) pattern were also recorded along with prominently observed d-value of different crystallographic planes. USB-2000 UV-visible spectrometer was used to measure the optical spectra of nanoparticles over the range 200 nm to 800 nm. The photoluminescence spectra were recorded in the wavelength range 300 nm to 700 nm with excitation wavelength 325 nm and 350 nm by an F-2500 Hitachi FL spectrometer. The FTIR spectra were recorded in the wave number range 4000–400 cm^{-1} using an FTIR spectrometer (IR affinity-1, Shimadzu, Japan). The temperature dependence of electrical resistivity and conductivity of ionic colloidal copper sulphide solutions was determined using a standard Conductivity Cell. A conductivity cell in a measuring solution is placed in the inverting input path of an operational amplifier. When AC voltage of constant amplitude and suitable frequency is applied to the system then for a given feedback resistance the output is linearly proportional to the conductance of the solution. On multiplying conductance by cell constant gives the ionic conductivity of the solution.

III. Result and Discussion

A. Structural Study

The structure of the synthesized nanoparticles was determined by XRD analysis. The XRD pattern shows prominent lattice planes (1 0 2), (1 0 3), (1 1 0), and (2 0 3) with 2θ at 29.51°, 32.09°, 48.23°, and 59.37° respectively corresponding to typical CuS (covellite) phase having hexagonal structure along with weak diffraction peaks superimposed on starch matrix. The calculated lattice parameters are (Table 1) in close agreement with standard data (ICSD: 061797) and also with reported results of others [19, 20]. The indexed lattice planes of selected area diffraction pattern (SAED) are also in agreement with the XRD results.The room temperature grown copper sulphide (S1) peaks are highly intense and broad compared to the high temperature synthesized CuS (S2) suggesting formation of good crystalline and mono-dispersed nanosizeCuS nanoparticles in sample S1. Table 1 shows the structural parameters of the synthesized product estimated from

Debye-Scherrer and Williamson-Hall plot. The Debye-Scherrer relation is given by the equation:

$$\beta = \frac{k\lambda}{DCos\theta} \text{ or } Cos\theta = \left(\frac{k\lambda}{D}\right)\frac{1}{\beta} \quad \text{... (1)}$$

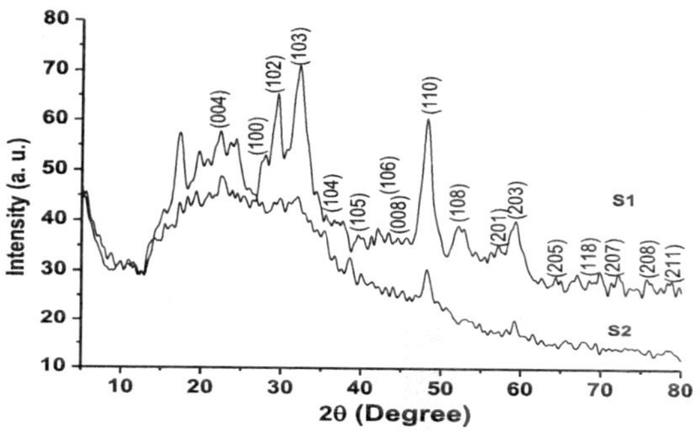

Figure 2: XRD Spectra of Samples S1 and S2 Synthesized at Room Temperature and Elevated Temperature Respectively

where k is shape factor ~0.9, λ (0.15406 nm) is the wavelength of Cu K_α radiation, D is the crystallite size, β is full width at half maximum (FWHM) of the most preferential peak in radians and θ is the peak position for the corresponding (hkl) [21]. From the slope of the plot $1/\beta - Cos\theta$, the average grain size (D) was calculated. The average crystallite size and strain

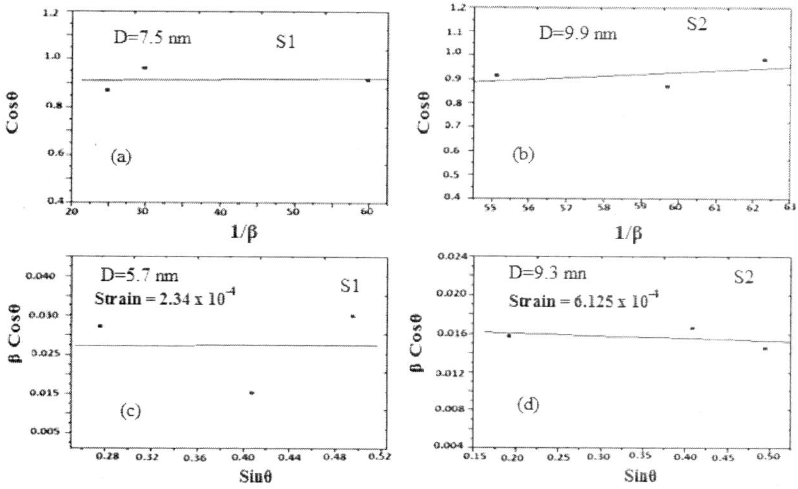

Figure 3: (a, b) Average Particle size (Debye-Scherrer plot), (c, d): Average Particle Size and Strain (W-H plot) of Samples S1 and S2

induced broadening arising out of crystal defects and distortions were also estimated by W-H plot from XRD data [22, 33]. The W-H relation includes a tensile strain ε in addition to Debye-Scherrer term as given by the equation:

$$\beta Cos\theta = \frac{k\lambda}{D} + 4\varepsilon Sin\theta \quad \text{... (2)}$$

The plot Sinθ−βcosθ gives grain size D and strain ε of the product. The results of Debye-Scherrer plot and W-H plot are in well agreement with HRTEM results. For sample S2, we obtained larger particle size compared to S1 which may be due to agglomeration of CuS nanoparticles at high temperature [15].

Table 1: Structural Parameters of the Synthesized CuS

Samples	*Molar conc./ Temperature*	*Lattice Constant (Å)*		*Particle Size (D) (nm)*		*Strain (ε)*
		a	*c*	*D-S*	*W-H*	
S1	0.1 M/30°C	3.77	16.03	7.5	5.7	2.34×10^{-4}
S2	0.04 M/85°C	3.79	16.12	9.9	9.3	6.125×10^{-4}

B. Morphology Study by HRTEM and SAED

The HRTEM morphology shows a good distribution of spindle type CuS nanoparticles (Figure 4(a)) of nearly spherical shape (Figure 4(d)) embedded in starch matrix while SAED pattern (Figure 4(c)) shows a central halo with concentric broad rings, the broadness, possibly originating from the confinement of nanocrystallites embedded in the starch matrix for the typical CuS synthesized by keeping volumetric ratio Cu:S = 1:2 and molarity 0.005 M [23]. The corresponding rings of SAED pattern may be indexed to (1 0 2), (1 0 3), (0 0 8), (1 0 8), and (2 1 1) crystal planes of hexagonal structure of CuS. The spindle particles have average dimensions of 100 nm that are actually the agglomerates of nanoparticles having diameter 13 nm (Figure 4(b) and 4(d)). It is observed that the capping agents play an important role in the confinement of spindle particles. The rings are prominent which indicates a good crystallinity possessed by the synthesized nanoparticles. The HRTEM also shows the d-value of 0.29 nm in spindle particles (Figure 2(e)) (Inset) corresponding to (1 0 2) plane indicating the growth direction along that atomic plane.

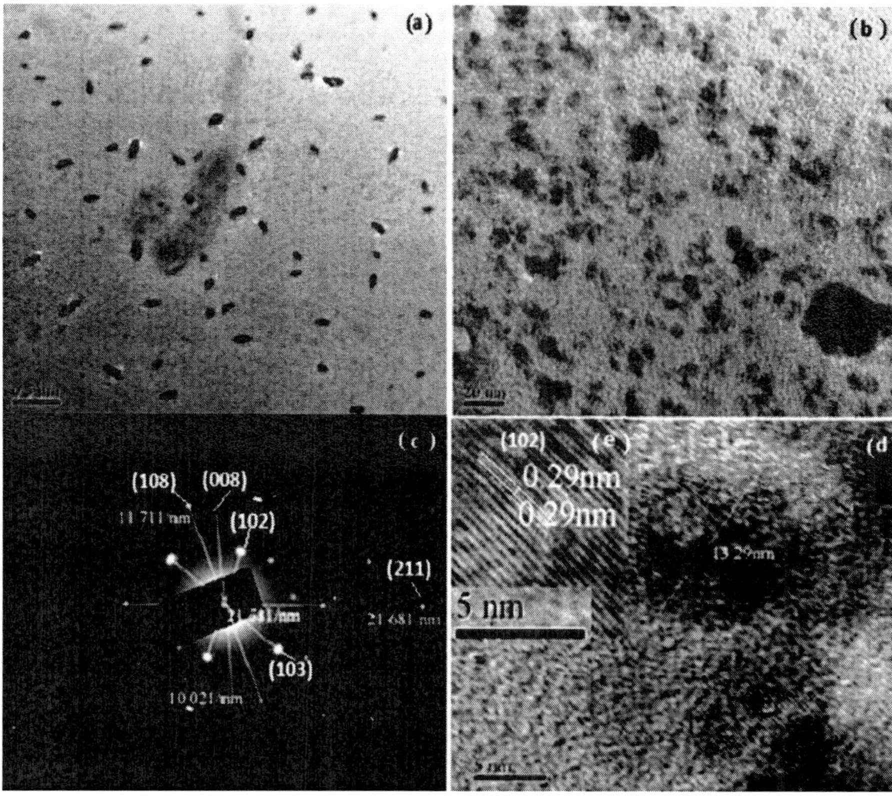

Figure 4: (a) HRTEM Images of Spindle Type Nanoparticles, (b) High Resolution of Spindle Particles (c) SAED Pattern (d) Size of a Nanoparticle (e) D-spacing (Sample S4)

C. Optical Study

The optical properties of the synthesized product were determined by UV-Vis absorption, PL and FTIR spectroscopy. The UV-Vis spectroscopy (Figure 5) show good blue shift of absorption edge for all synthesized materials (Table 2). The band gap is enhanced to 3.31–3.73 eV on lowering molar concentration. The optical band gap was determined by Tauc- plot $(\alpha h\nu)^2$-hν plot. The linearity of the plot indicates that the mode of transition is of direct nature [24]. The spectra show a wide absorption in the near IR region characteristic of hexagonal structure of covellite phase of CuS [25]. Figure 5(h) exhibits an increase of band gap with lowering molarities which is in well agreement with others [11].

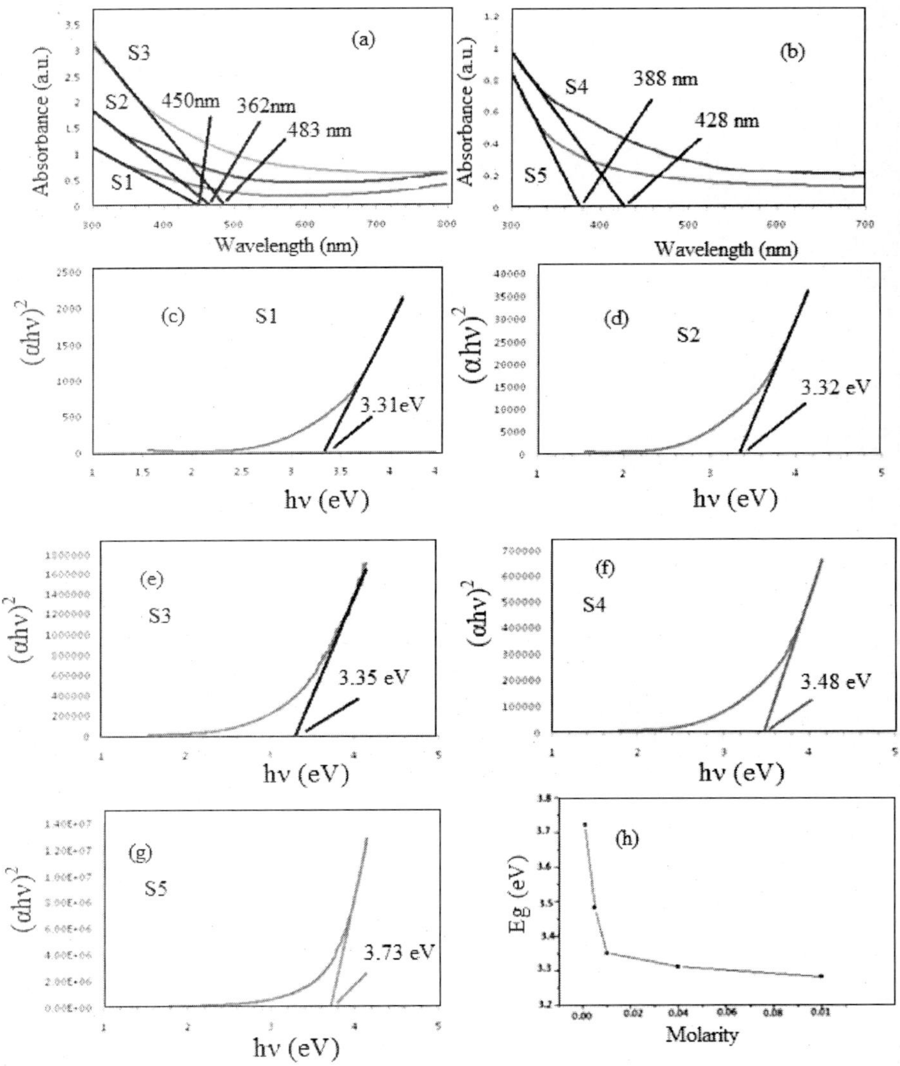

Figure 5: (a, b) UV-Visible Spectra of Samples S1, S2, S3, S4 and S5, (c, d, e, f, g): Corresponding Tauc plot, (h): Molar Con.-Band Gap Plot

The PL spectra show near UV emissions around 362 nm to 395 nm measured at excitation wavelength 325 nm (Figure 6, S4, S6) and another extrinsic emission at 655 nm (Figure 6, S1). However, near band gap emissions were not observed in two samples (Figure 6, S3, S7) measured at excitation wavelength 350 nm where only extrinsic emissions were observed around 705 nm. The observed near band gap emission appears to be as a result of radiative recombination of electrons and holes in confinement of CuS nanoparticles in the highly reactive surface of atoms whereas the extrinsic emission may arise due to various defects in CuS nanocrystals which is apparent from the calculated high value of stokes shift (Table 2). The sharp emission peaks can be attributed to the quantum confined enhanced band gap emissions which become strong on lowering molar concentration. The strong band gap emission demonstrates the high crystalline nature of the as-synthesized spindle particles [26]. The extended of PL peaks and their intensity enhancement may be attributed to the good aspect ratio of nanoparticles. A good aspect ratio multiplies and increases the surface to volume ratio that induces large surface states. These surface states contribute to the enhancement of photoluminescence intensity as well as extend the emission further in the long wavelength range. This PL distribution in the near UV-blue region is originated from the uniform distribution of well defined spindle particles [27, 34]. Hence, nanostructure CuS synthesized at such lower molar concentration may be fruitfully used in the fabrication of UV-blue LED.

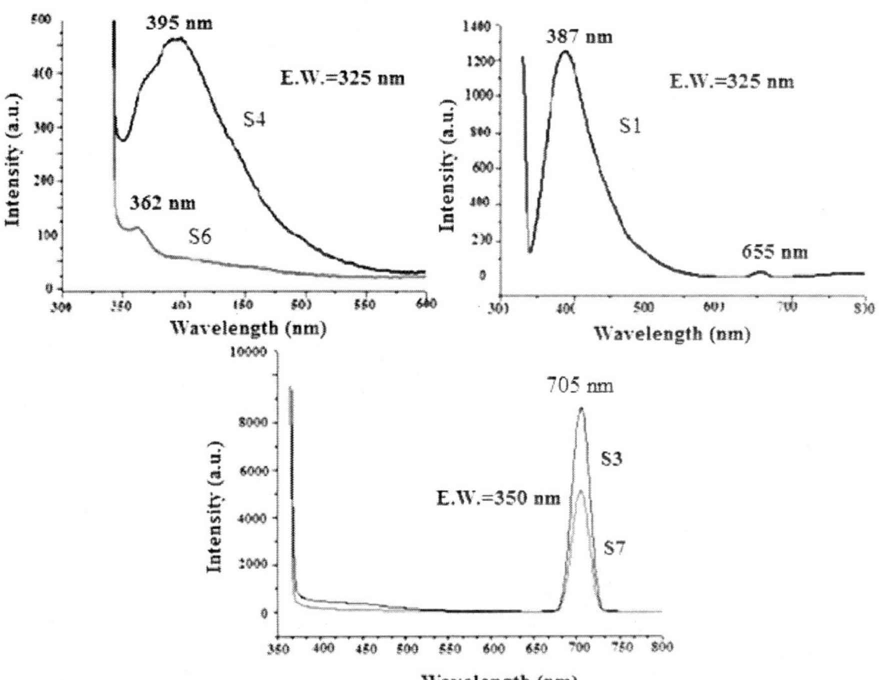

Figure 6: PL Spectra of Samples S1, S3, S4, S6 and S7

Table 2: Optical Properties of Starch Capped CuS: PL Peaks and Band Gaps and Stokes Shifts

Molar Con./ Samples	*Excitation Wavelength*	*PL Emission (nm)*		*UV-Vis*		*Stokes Shifts (nm)*
		Near Band Gap Emission (λ_e)	*Impurity Emission (λ_i)*	*Absorption edge (nm)*	*Band Gap (eV)*	
0.1 M/S1	325	387	665	450	3.31	53
0.04 M/S2	350	....	...	462	3.32	...
0.01 M/S3	350	...	705	483	3.35	...
0.005 M/S4	325	395	...	388	3.48	18
0.001 M/S5	325	...	....	428	3.73	..
0.05 M/S6	325	362	....	....	...	..
0.07 M/S7	350	...	705	....	...	..

D. FTIR Study

Figure 7 shows FTIR spectra of various vibration modes of pure starch and starch capped CuS located at different regions. The absorption peak of –OH group of starch (S1) of CuS located at 3435 cm^{-1} is slightly shifted from that of pure starch (S8) located at 3419 cm^{-1}. This indicates a strong interaction between CuS nanoparticles and –OH group of starch. Thus, CuS nanoparticles and –OH group of starch may be bound by chemical interaction which prevents CuS nanoparticles from aggregation in nanocrystals [28]. The absorption band at 615 cm^{-1} is attributed to Cu-S vibration band of CuS nanoparticles which matches well with results of others [29]. The peak located around 1604 cm^{-1} is attributed to C=O stretching mode while strong peak around 1014 cm^{-1} may be due to the –C–O group of starch indicating CuS nanoparticles are well confined in starch matrix.

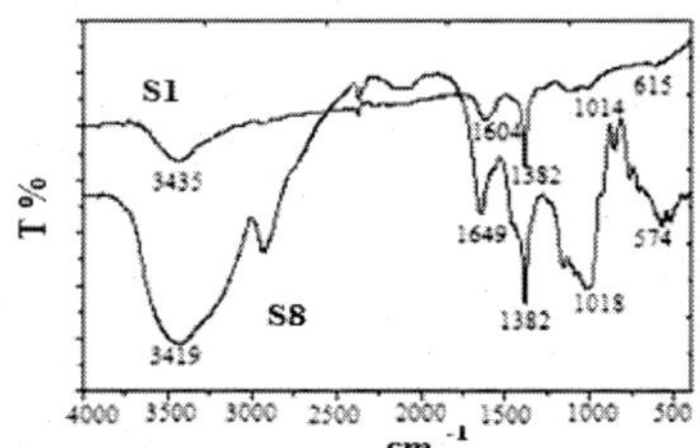

Figure 7: FTIR Spectra of Pure Starch (S8) and Starch Capped CuS Nanoparticles (S1)

Table 3: Functional Groups Present in Starch Capped CuS

Vibrational Mode	***Absorption (cm^{-1}) Region***	***Relative Intensity***
Cu–S stretching	400–650	Small and weak
C=O stretching	1604	Medium
–C–O stretching	1014	Medium and broad
O–H stretching	3435	Medium and broad

E. Temperature Dependence of Ionic Conductivity of CuS Nanoparticles

The electrical conductivity variation of ionic CuS nanoparticles solution with temperature in temperature range 281–359 (0.1 M solution) and 283–359 K (0.04 M solution) were measured using a standard conductivity cell. From the measured conductivity the resistivity of ionic solution at different temperature was calculated. The resistivity vs. inverse temperature plots well exhibit the semiconducting nature of CuS nanoparticles and are in good agreement with the reported results [30] (Figure 6(a, b)).The ionic conductivity of CuS ionic solution is given by the simple exponential relation

$$\sigma_i = \sigma_{i0} \exp(-E_a/k_B T)$$

or,

$$\text{Resistivity, } \rho = \rho_0 \exp(E_a/k_B T) \qquad \text{... (3)}$$

where σ_{i0} and ρ_0 are pre-exponential factors, E_a is activation energy for the conduction process of copper ions, T is absolute temperature of ionic solution [16]. The activation energy was calculated from the slope of the resistivity vs. inverse temperature plot using the relation

$$E_a = 2.303\ k_B \text{slope (eV)} \qquad \text{... (4)}$$

where k_B = Boltzmann constant = 8.617×10^{-5}eV/K.

The details results are given in Table 4. The activation energy for both the samples is quite in agreement with reported activation energy values [31, 32]. Molar conductivity, degree of dissociation (α), solubility product K_{SP} and ionic mobility (μ) of the ionic CuS solution were also calculated from conductivity data at 30°C as listed in Table 4. As the degree of ionization increased with increasing dilution the molar conductivity also increased which is the basis of the theory of an ionic solution. The degree of dissociation is less than 1 further indicating week electrolyte nature of ionic CuS solution. Solubility decreased with dilution which is also the theoretical basis of an ionic solution.

Table 4: Electrical Properties of Ionic CuS Solution

Samples /Molar Con.	σ $(\Omega^{-1}cm^{-1})$ $(30^{\circ}C)$	ρ $(\Omega\ cm)$ $(30^{\circ}C)$	E_A (eV)	$\wedge$ $(\Omega^{-1}cm^{2}$ $mol.^{-1})$	α	K_{SP}	μ $(cm^{2}$ $s^{-1}V^{-1})$
S9/0.1M	7.2081×10^{-3}	1.3873×10^{2}	0.1248	72.0814	0.058 (5.8%)	100×10^{-5}	23.5×10^{-2}
S10 /0.04M	3.05×10^{-3}	3.278×10^{2}	0.1602	76.2931	0.061 (6.1%)	1.6×10^{-5}	

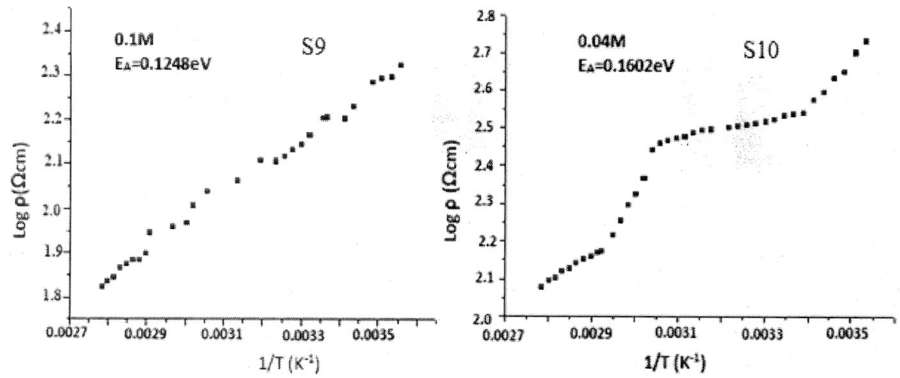

Figure 8: Resistivity – Temperature Plot of Samples S9 and S10

IV. Conclusion

A spindle type particle distribution was obtained from the structural morphology of CuS nanoparticles synthesized through chemical route. A simple chemical route was employed which is well regarded as nontoxic, environment friendly and requires low cost infrastructure. The starch as capping agent played important role in controlling the structural morphologies as well as in confinement of nanoparticles. Thiourea was used as sulphur source because release of S^{2-} ions from thiourea in aqueous ammonia can be controlled by adjusting the pH value of the reaction. The XRD measurement showed covellite phases of CuS superimposed on the product materials. HRTEM morphology exhibits well defined spindle type nanoparticles having average size 100 nm. It is observed from the images that the spindle particles seems to be results of diffusion induced agglomerisation of some individual spherical particles of average diameter 13 nm. The optical properties of the CuS nanoparticles exhibited good blue shift of absorption edge with enhancement of band gap from 3.31 eV to 3.73 eV with lowering of molarities. The PL spectra show a non-symmetric emission around 362–395 nm. This UV-blue emission corresponds to the spindle particle distribution and hence the intensity increases because of their good aspect ratio synthesized at lower molar concentration. The ionic conductivity study approach in an electrolytic ionic solution reported here virtually with no reported papers is found to be suitable in determining the semiconducting nature of CuS nanoparticles. The temperature dependence of electrical resistivity variation of the ionic solution shows that resistivity value decreases with temperature confirming semiconducting nature of the synthesized CuS nanoparticles. The experiment reveals that the results can be utilized for fluorescence based devices such as UV-blue LED exploiting nanosize CuS synthesized at lower molarities.

Acknowledgment

The author would like to acknowledge the technical support of the Department of Chemistry, Gauhati University for UV-Vis and PL measurements, SAIF, Gauhati University for XRD analysis, SAIF, North Eastern Hill University for HRTEM measurements and Department of Chemistry, B. H. College for the ionic conductivity measurements.

References

[1] J. Zou, J. Zhang, B. Zhang, P. Zhao, X. Xu, J. Chen and K. Huang (2007). *Mater Sci.,* 42, 9181.

[2] K.V. Singh, A.A. Martinez-Morales, G.T. Senthil Andavan, K.N. Bozilov and M. Ozkan (2007). *Chem. Mater,* 19, 2446.

[3] J.P. Tailor (2014). CuS-Single Crystals, Thin Films, Nanoparticles and their application (Sardar Patel University, Gujrat, 2014, Thesis).

[4] M. Annie Freeda, N. Rode Madhav, C.K. Mahadevan and S. Ramalingom (2010). *Nanotech. and Nanosc.* 1, 4.

[5] F. Zhuge, Li, X. Gao, X. Gan and F. Zhou (2009). *Materials Letters,* 63, 652.

[6] J.C.W. Folmer and F.J. Jellinek (1980). Less-Common Metals, 76, 153.

[7] W. Liang, and M.H. Whangbo (1993). Solid State Commun. 85, 405.

[8] U.K. Gautam and B. Mukherjee (2006). *Bull. Mater. Sci.,* 29, 1.

[9] P. Lukashev and W.R.L. Lambrecht (2007). *Phy. Rev.,* 76, 195202.

[10] L.A. Isac, A. Duta, A. Kriza, I.A. Enesca and M. Nanu (2007). *J. of Phy. Conference Series,* 61, 477.

[11] M.A. Sangamesha, K. Pushapalatha and G.L. Shekar (2013). *IJRET*, 2, 2321.
[12] I. Popovici, L. Isac and A. Duta (2009). *Bulletin T. Uni. Brasov,* 2, 193.
[13] A. Kolodziejczak-radzimska and T. Jesionowski (2014). *Materials (Basel),* 7, 2833.
[14] H.T. Boey, W.L. Tan, N.H. Abu Bakar, M. Abu Bakar and J. Ismail (2007). *J., Phy. Sci.* 18, 87.
[15] Zhang, J.Z. (2000). *J. Phys. Chem.,* 104, 7239.
[16] K. Okamoto and S. Kawai (1972). *Japanese J. Appl. Phy.,* 12, 1130.
[17] J.A. Shetzline and S.E. Creager (2014). *J. Electrochem. Soc.*, 161, 917.
[18] Zhuge, F. Li, X. Gao, X. Gan, X. and F. Zhou (2009). *Materials Letters,* 63, 652.
[19] A. Phuruangrat, T. Thongtem and S. Thongtem (2011). *Chalcogenide Lett.,* 8, 291.
[20] K. Yao, W. Lu, X. Li and J. Wang (2012). *Solid State Chem.,* 196, 557.
[21] J.P. Tailor, A.J. Khimani and S.H. Chaki (2018). *AIP Con. Proc.,* 1961, 030012.
[22] T.M.K. Thandavan, S.M.A. Gani, C.S. Wong and R. Nor (2015). *J. Nondestruct Eval,* 34, 13.
[23] R. John and S.S. Florence (2010). *Chalcogenide Letters,* 7, 269.
[24] S.S. Dhasade, J.S. Patil, J.S. Thombare and V.J. Fulari (2015). *J. of Shivaji Univ. (Sc. & Tech.),* 41, 2.
[25] S.H. Chaki, J.P. Tailor and M.P. Deshpande (2014). *Material Sc. In Semicond. Proc.*, 27, 577.
[26] G. Carotenuto, F. Caoezzto, M. Palomba and F. Nicolais (2010). *Int. J. Nansc.*, 9, 385.
[27] S.N. Sahu and K.K. Nanda (2001). *PINSA,* 62, 103.
[28] J. Xu, X. Cui, J. Zhang, H. Liang, H. Wang and J. Li (2008). *Bull Mater,* 31, 189.
[29] M. Saranya, C. Santhosh, R. Ramachandran and A.N. Grace (2014). *J. Nanotech.,* 2014, 7.
[30] S.L. Prokopenko, G.M. Gunja, S.N. Makhno and P.P. Gorbyk (2015). *Chem. Phys. Tech. Surface,* 6, 234.
[31] I.M. Sung Soon, I.M. Hee Suk and Y.K. Eun (1990). *J. Appl. Polymer Sc.,* 41, 1517.
[32] J.S. Cruz, S.A.M. Hernandez, F.P. Delgado, O.Z. Angel, O.Z., R.C. Perez and G.T. Delgado (2013). *Int. J. Photometry,* 2013, 7.
[33] D. Nath, F. Singh and R. Das (2020). *Mater. Chem. Phys.*, 239, 1.
[34] B. Barman and K.C. Sarma (2020). *Solid State Sci.,* 109, 1.

CHAPTER 5

Green Photovoltaics: The Promise of Natural Dye-Sensitized Solar Cells

Smritimala Sarmah*, Hena Choudhury, Evlina Kalita, Dhritisman Krishnatreya and Gitartha Sarma

Department of Physics, B. Borooah College, Guwahati, Assam, India

*E-mail: sarmahsmritimala15@gmail.com

ABSTRACT: Natural dye-sensitized solar cells, or NDSSCs, have become a viable substitute for conventional silicon-based photovoltaics because of their easy manufacture, affordability, and environmental sustainability. The photo-sensitization method, which is the basis for these solar cells, involves organic dyes sensitising broad bandgap semiconductors, most frequently titanium dioxide (TiO_2). By absorbing visible light, these dyes allow electrons to be injected into the conduction band of semiconductors. Natural colours have garnered a lot of attention lately, especially those produced from plants like betalains, carotenoids, anthocyanins, and chlorophylls. These colours' availability, non-toxicity, and biodegradability make them attractive. The performance, stability, and effectiveness of NDSSCs are examined in this work using a variety of natural dyes that are taken from fruits, vegetables, and leaves.

Important factors such as dye adsorption, the light absorption spectrum, and charge transfer kinetics were examined to determine how they affected the solar cells' overall efficiency. Despite having lower power conversion efficiencies (PCEs) than their synthetic counterparts (usually between 0.2% and 2.0%), natural dyes offer important information for solar energy solutions that are sustainable. The study critically examines variables that affect NDSSC performance, including co-sensitization techniques, dye purity, and dye solution pH.

Thus, while natural dye-sensitized solar cells currently face efficiency challenges, ongoing research into dye chemistry, materials engineering, and device optimization may pave the way for NDSSCs to become viable components in next-generation sustainable energy systems. The integration of eco-friendly materials and green chemistry approaches positions NDSSCs as a promising area for further innovation and development.

Keywords: Natural Dye Sensitized Solar Cells, Green Photovoltaics, Sustainable Energy Systems, Titanium Dioxide, Plant-Based Dyes.

I. Introduction to Photovoltaics

A. The Need for Sustainable Energy

Traditional sources of energy like fossil fuels are limited and their use also cause environmental issues like global warming, climate change, etc. as well as it also affects the human health. Therefore, over the last few decades studies and research have been conducted to find an environmentally friendly renewable sources of energy that can fulfil the increasing global demand.

In recent decades, research has increasingly concentrated on evaluating the detrimental effects of global warming and the significant rise in greenhouse gas emissions on everyday life. This human-induced phenomenon has profoundly disrupted climate patterns, adversely impacting both public health and economic development. Reports have notably indicated that environmental conditions, food production, and water accessibility are critically compromised, leading to existing damages with potential for exacerbation in the future. Medical research has linked worsening cases of diarrhoea, dengue fever, and malaria to climate change. Additionally, respiratory issues and cardiovascular health concerns are rising due to the increased frequency of heatwaves and air pollution caused by various industrial activities, including power generation. Furthermore, alterations in the ozone layer and rising sea levels have also influenced agricultural sectors, indirectly compromising human health through poor nutrition [1]. There are many obstacles to overcome while switching from fossil fuels to renewable energy sources. Studies at the laboratory and industrial levels have shown how difficult it is to strike a compromise between crucial elements such as preserving low or non-polluting energy sources while guaranteeing energy efficiency. Therefore, lowering carbon dioxide emissions together with other air pollutants like sulphur dioxide, nitrogen oxides, etc. that severely impair air quality and have an adverse effect on human health and other living things worldwide is a major objective in this transition. Around the world, several regulations are already in place to help transition from fossil fuels to renewable energy sources in different amounts over the ensuing years. However, the development of renewable energy technologies must consider their capacity for generation, installation, and storage [1].

Subsidies are being provided by governments worldwide to promote the use of renewable energy sources. Recently, startups and businesses engaged in wind and solar photovoltaics have received a lot of support. The viability of these alternative energy sources considering regional climates and geographic locations is a crucial factor. For example, wind energy is steadily gaining traction in Latin America and Africa, whereas solar thermal energy seems more feasible in North Africa, South Africa, and the Middle East. Numerous initiatives, including progressive policies, funding opportunities, and enhanced technical help, are supporting this rise. These initiatives are expected to pave the way for a sustainable future while assisting in minimizing current harm. It is essential to conduct a thorough evaluation of the total effects of energy systems to precisely determine the health repercussions related to each of their phases of life, from the extraction of raw materials to ultimate consumption. Generally speaking, health risks fall into three categories: first, fuels made from biomass or fossil fuels like coal or natural gas, which result in substantial solid waste and air pollution; second, low-energy-density renewable energies (like solar or wind), which usually produce few emissions except in the event of an accident; and third, nuclear power, which has high energy densities and requires less fuel input while producing little waste for transformation. Based on varying emission rates across each energy source category, this classification makes it easier to evaluate each one's effects and related toxicity levels [1].

Solar energy stands out as one of the most practical and promising renewable power sources available today. Over the years, numerous technologies have emerged that demonstrate its versatility across many uses—ranging from providing heat and cooling to powering dryers, generating electricity, and even purifying water through desalination. Among these, solar photovoltaic (PV) technology has become a key player in the global shift toward sustainable energy. It converts sunlight directly into electricity through semiconducting materials that exhibit the photovoltaic effect. This process forms the foundation of many clean energy innovations. Most PV modules are made using silicon-based cells, which enable the conversion of sunlight into electrical power by freeing electrons from their atomic bonds when struck by photons. The growth of solar PV has been rapid and remarkable. By 2022, it achieved a production milestone of about 270 terawatt-hours—a 26% increase from the previous year. This surge allowed solar PV to overtake wind energy for the first time in history in terms of annual electricity generation growth. Today, solar PV stands as the third-largest source of renewable energy worldwide, providing roughly 4.5% of the total global electricity output, behind hydropower and wind energy [2].

B. Basic Principles of Solar Energy Conversion

The photovoltaic effect is the basic physical process by which a photovoltaic cell converts solar energy into electrical energy. Sunlight is formed by photons, the concentrated packets of energy emitted by the sun [3]. These photons' wavelengths in the solar spectrum affect their energy. Photons can either be absorbed or reflected when they encounter a photovoltaic cell. When a photon is absorbed, its energy is transferred to an electron in the solar cell, typically within a silicon atom allowing electricity to be produced [4]. The electron can escape from its normal position inside the atom and join the current in an electrical circuit. Manufacturers combine two different types of semiconductors (p-type and n-type) to create a junction that produces the electric field inside a solar cell, as illustrated in Figure 1. The most popular method for creating p-type or n-type silicon material is to add an element with an extra electron or an electron deficiency [5]. The bandgap energy of the semiconductor material determines a photon's capacity to excite an electron. An electron is excited from the valence band to the conduction band when a photon transmits sufficient energy, resulting in the creation of an electron-hole pair [6]. These charge carriers are then crucially separated by the p-n junction's intrinsic electric field. It creates an electric potential difference by directing free electrons toward the n-type area and holes into the p-type region [7]. When the circuit is closed, this electron movement causes an electrical current to occur. However, efficiency may be decreased if certain charge carriers recombine before contributing to the production of power. The most common material used in the production of solar cells is silicon. It can absorb photons with energy equal to or higher than this threshold because of its bandgap, which is roughly 1.1 eV [8].

II. Limitations of Conventional Solar Cells

A. Efficiency and Cost

Crystalline silicon is the most common photovoltaic material. As the most developed photovoltaic material and the long-term market leader, crystalline silicon is widely available, non-toxic, inexpensive, and enables the creation of cells with high and consistent conversion efficiency. Both the integrated circuit and photovoltaic sectors have extensive and profound infrastructure and expertise in crystalline silicon technology.

Under non-concentrating sunlight, the crystalline silicon solar cell's theoretical limiting efficiency is roughly 29%. This is not far below any single junction solar cell's theoretical limit. Under non-concentrated sunlight, the best experimental and commercial silicon solar cells currently achieve 24–25% efficiency, or roughly 85% of the theoretical limit [3].

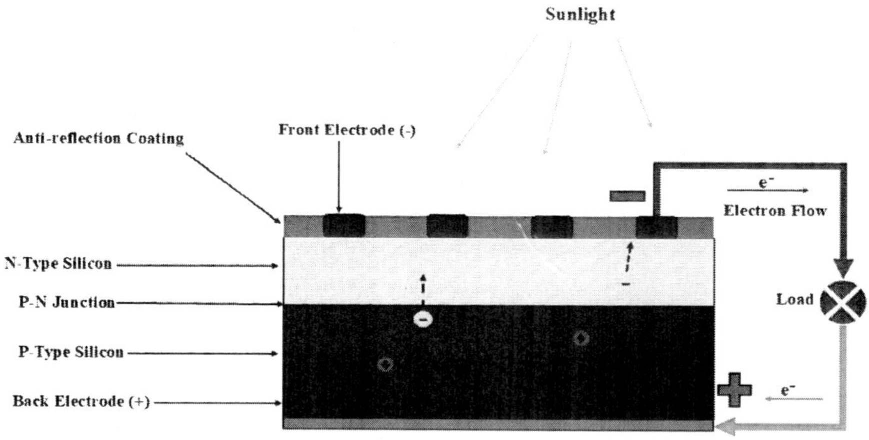

Figure 1: Schematic Diagram of Photovoltaic Cell

The passivated emitter with rear locally diffuse cell, shortly known as PERL cell, which is based on p-type silicon, reached a record for energy conversion efficiency of 25% in 1999 and stayed unmatched for 15 years. Thanks to heterojunction (HJ) intrinsic thin layer technology based on thin amorphous silicon (a-Si) passivating layers and interdigitated back contacts (IBCs) on n-type silicon wafers, the record efficiency increased to 25.6% in 2014 and to 26.7% in 2017. Wafer-based silicon modules have an efficiency of 24.4% and are continuously improving in both the lab and the market [4].

Crystalline silicon accounts for 90% of the global photovoltaics (PV) market and has grown at a nearly 30% annual rate, although accounting for less than 2% of total power capacity. To maintain this growth trajectory, further cost and capital expenditure (capex) reductions are required. Thinning silicon wafers below the industry standard of 160 μm decreases manufacturing costs and capex, accelerating the economically sustainable expansion of PV manufacturing [5]. Cost is a significant element in photovoltaic technology adoption. Current commercial silicon modules cost between $0.20 and $0.30 per watt, including silicon purification, wafer slicing, cell fabrication, and module assembly [6].

Dye-sensitized solar cells (DSSCs) are a type of photovoltaic device that generate electricity by using a light-absorbing dye to capture sunlight and transfer energy to a semiconductor, typically TiO_2, enabling electron flow and power generation. DSSCs are considered more cost-effective mainly because they rely on inexpensive materials and simpler manufacturing processes compared to silicon-based solar cells.

Firstly, DSSCs use cheap and abundant materials. The photoanode is typically made from TiO_2, which is low-cost, widely available, and easy to process. The dyes used to absorb sunlight are also relatively inexpensive compared to the high-purity silicon required for conventional solar cells. Secondly, silicon solar cells require costly processing. Producing silicon-based cells involves high-temperature purification, crystal growth, and complex fabrication steps, all of which significantly increase manufacturing costs. In contrast, DSSCs can be fabricated at low temperatures using simple techniques such as screen printing or coating, reducing energy and equipment expenses. Finally, DSSCs do not require ultra-clean manufacturing environments, further lowering production costs. Because of these factors DSSCs are generally more cost-effective than traditional silicon-based solar cells, especially for low-power or flexible applications.

B. Environmental Concerns in Production and Disposal

Photovoltaic (PV) systems are thought to be sustainable and clean energy sources. Even though PV systems operate with very little pollution throughout the course of their lifetime, it is impossible to overlook the potential environmental effects of these systems from production to disposal. Important environmental factors include harmful waste creation, water pollution, air emissions from manufacturing, and land-use impacts of PV systems. In addition to providing innovative design suggestions to lessen and resolve the environmental issues, the current study attempts to offer a thorough analysis of all potential environmental concerns. Greenhouse gas (GHG) emissions from different PV systems were also investigated and contrasted with energy sources derived from fossil fuels. The findings showed that optimal design, the creation of new materials, reducing the use of hazardous components, recycling whenever feasible, and careful site selection could significantly reduce the adverse environmental effects of PV systems. These mitigation measures will lower greenhouse gas emissions into the atmosphere, lessen the buildup of solid waste, and protect important water resources.

It was discovered that the carbon footprint emissions from PV systems ranged from 14 to 73 g CO_2-eq/kWh, which is 10 to 53 orders of magnitude less than the emissions from burning oil (742 g CO_2-eq/kWh). It was determined that the use of innovative manufacturing materials might reduce the PV system's carbon footprint by an additional order of magnitude. Additionally, recycling solar cell components can help cut greenhouse gas emissions by up to 42%. The current study provides a useful management approach that can be applied to enhance PV manufacturing processes' sustainability, increase their economic worth, and lessen their detrimental effects on the environment [7].

Solar panels typically last 25–30 years before becoming useless. Some solar panels include hazardous materials such as lead, cadmium, selenium, and other toxic metals, which can contaminate soil and groundwater if disposed of improperly, such as by landfilling. Solar panel trash is expected to total roughly 78 million tons globally by 2050, posing both an environmental issue and an economic opportunity. Without adequate management, these materials can harm ecosystems and pose health hazards.

III. The Emergence of Dye Sensitized Solar Cell (DSSC)

A. History and Invention of Gratzel Cell

People are turning away from conventional energy sources like coal and nuclear power plants as fossil resources run out and the price of crude oil continues to rise. People are becoming more aware of renewable energy sources as a result. Hydropower, wind power, geothermal power, and the conversion of biomass into energy are a few instances of renewable resources. Solar thermal collectors and photovoltaics are two more significant renewable energy sources. When it comes to DSSCs, natural dyes have become an affordable and environmentally beneficial substitute for synthetic dyes. The quest to convert solar radiation into useful energy is as old as science itself. Prior to silicon solar panels taking over rooftops, scientists worldwide searched for innovative ways to replicate photosynthesis, which is a process found in nature. Among the most significant breakthroughs in this quest stands the dye-sensitized solar cell, widely known as the Gratzel cell, after its co-inventor Michael Gratzel. The journey to this invention is not just about materials and molecules; it is a story of steady curiosity, teamwork, and humanity's awe of nature's light-driven processes.

The roots of DSSC science stretch back to the late 19th and early 20th centuries, when scientists first noticed that organic dyes, when exposed to light, could drive tiny electric currents on certain electrode surfaces. Yet, for a long while, these observations had no real application—after all, the currents produced were minuscule and unreliable.

A new group of scientists investigated the idea again in the 1960s and 1970s, when research into solar energy was heating up. Researchers wondered if we could use dyes like chlorophyll, which plants use to get energy from sunlight, to collect light and create useful currents in man-made devices. Researchers at the University of California, Berkeley, started to "sensitize" zinc oxide (ZnO) and TiO_2 electrodes with chlorophyll that had been taken from plants. They used a dye-sensitized ZnO electrode to convert light to electricity for the first time in 1972. What is the basic idea? A photon of light would excite a dye molecule, which would then give an electron to the semiconductor electrode, making a current [9]. The efficiency of these DSSC's is very poor because of limited absorption, low stability of dyes, as they frequently decomposed under prolonged illumination, and poor electron collection.

Everything changed in the 1980s, when scientists began to explore the properties of nanostructured materials. Instead of using flat electrodes, electrode out of a layer of extremely fine nanoparticles, almost like a sponge, which vastly increasing its surface area are used. Such a design would allow thousands of times more dye molecules to be attached, multiplying the light absorbed. The real breakthrough came when Michael Gratzel, a Swiss chemist at the Ecole Polytechnique Federale de Lausanne (EPFL), and his then-postdoctoral collaborator, Brian O'Regan, combined the latest insights in nanomaterials with their interest in light-harvesting dyes. Motivated in part by the oil crisis and the promise of clean, affordable solar technology, Gratzel's team pursued a radical new architecture for dye solar cells [9].

B. Structure and Working Principle of DSSCs

DSSCs are cutting-edge technology that replicates natural photosynthesis to turn sunlight into electricity. Their separation of the roles of charge transport and light absorption sets them apart from conventional silicon solar cells. Four essential components of a DSSC are the counter electrode, redox-mediator (electrolyte), sensitizer (dye), and working electrode. To prevent electrolyte leakage, a DSSC consists of a working electrode coated with a sensitizer dye and sealed to a counter electrode containing a thin electrolyte layer using hot-melt tape. Let us examine their composition and operation in a thorough, understandable, and human-centred manner.

(a) *Structure of DSSCs*

Transparent Conductive Substrate: In a DSSC, the transparent conductive substrate serves as the front electrode, playing a dual role of allowing light to enter the cell and collecting electrons generated in the photoanode. It is typically made of glass or flexible plastic coated with a transparent conducting oxide (TCO), such as fluorine-doped tin oxide (FTO) or indium tin oxide (ITO). FTO is commonly used due to its high thermal and chemical stability, while ITO offers lower resistivity but is less stable at high temperatures. The substrate combines high optical transparency, which permits sunlight to reach the dye-sensitized semiconductor layer, with good electrical conductivity, which enables the collected electrons to flow to the external circuit. Thus, the transparent conductive substrate functions as both a window for light and a pathway for electron transport in the DSSC [9].

Working electrode (WE): A transparent conducting glass plate composed of FTO or ITO is coated with a thin layer of oxide semiconductor materials, such as TiO_2, Nb_2O_5, ZnO, SnO_2 (n-type), and NiO (p-type), to create the working electrode, which is around 10 μm thick. A dense layer of dye molecules may stick to the surface of this porous, sponge-like layer because of its extraordinarily large surface area, which is thousands of times greater than that of a flat surface. Due to its wide bandgap (3-3.2 eV), TiO_2 primarily transmits visible light and absorbs primarily ultraviolet light, allowing the dye to assume the primary role of absorbing sunlight. TiO_2 is primarily utilized as a semiconducting layer because it is non-toxic, cost-effective, and readily available, unlike alternative wide band gap oxides like ZnO and Nb_2O_5. These working electrodes are then submerged in a solution of a solvent and a photosensitive molecular sensitizer because these semiconducting layers only absorb a small portion of light in the ultraviolet spectrum. The dye is covalently bound to the TiO_2 surface after the film has been soaked in the dye solution. The huge surface area of the electrode and its highly porous shape cause many dye molecules to adhere to the nanocrystalline TiO_2 surface, increasing light absorption at the semiconductor surface [9].

Photosensitizer or Dye: The part of a DSSC that oversees maximizing the absorption of incident sunlight is dye. The light absorbers (photosensitizers) are the dye molecules affixed to the TiO_2 layer. Natural dyes, perovskite-based sensitizers, metal-free organic dyes, moderate dyes, ruthenium - complex dyes, and QD (Quantum dot) sensitizers are the different categories of photosensitizers [10]. Because of their strong charge transfer absorption throughout the visible spectrum and extremely effective metal to ligand charge transfer, ruthenium-based complexes are thought to be good sensitizers for DSSCs. However, due to their low cost, non-toxicity, environmental friendliness, ease of biodegradation, and above all large absorption coefficient, natural dyes have several advantages over metal complexes [11]. A sensitizer should contain carboxyl and hydroxyl groups, which is necessary for good semiconductor binding. Co-absorbents such chenodeoxycholic acid (CDCA) and anchoring groups like alkoxy-silyl phosphoric acid and carboxylic acid group were added between the dye and TiO_2 to prevent the dye from aggregating across the surface [9, 10].

Electrolyte: A redox electrolyte, usually a liquid solution containing iodide and triiodide, fills the space between the anode and the cathode. It regenerates the dye after electron injection and transports charges within the cell. Light weight, a high boiling point, low vapor weight, and good dielectric qualities are all desirable in an electrolyte. The ideal electrolyte solvent has low viscosity, low vapor pressure, a high boiling point, and exceptional dielectric qualities [10]. Liquid electrolytes, solid state electrolytes, and quasi solid-state electrolytes are the three categories into which DSSC electrolytes fall. For solid-state DSSC applications, iodide/triiodide incorporated into a polyaniline/thiourea matrix is also employed. Depending on the solvent used, liquid electrolytes can be broadly divided into two categories: room temperature ionic liquid electrolytes (RTIL) and organic solvent-based electrolytes [10].

Counter Electrode (CE): Usually, a thin layer of platinum (Pt) or carbon (C) catalyst is coated to a glass substrate as shown in Fig 2. The working and counter electrodes are then sealed together, and an electrolyte is added with the help of a syringe. The electrical circuit is completed by the counter electrode, which helps reduce triiodide back to iodide.

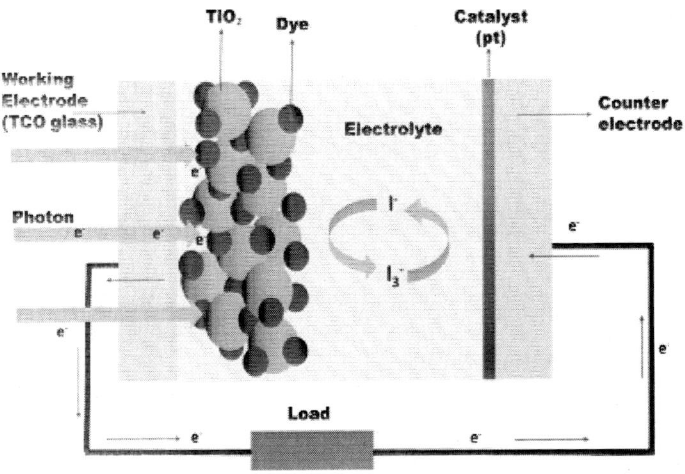

Figure 2: Schematic Diagram of the Working Principle of a Dye Sensitized Solar Cell (DSSC) Showing Photon Absorption, Electron Flow and Redox Reaction in the Electrolyte

(b) *Working Principle of DSSC*

DSSCs operate by converting sunlight into electricity through a series of photoelectrochemical processes. The working principle can be explained in the following steps:

- **Light Absorption and Dye Excitation:** The dye molecules coated on the semiconductor (usually TiO_2) absorb photons from sunlight. This absorption excites the dye molecules, promoting electrons from the ground state to a higher-energy excited state.
- **Electron Injection into TiO_2:** The excited electrons from the dye are injected into the conduction band of the TiO_2 semiconductor. This step generates free electrons in the TiO_2 while the dye becomes oxidized.
- **Electron Transport and External Circuit:** The injected electrons move through the TiO_2 layer to the transparent conductive substrate and then flow through the external circuit, producing electrical current, before reaching the counter electrode.
- **Dye Regeneration and Electrolyte Movement:** The oxidized dye molecules are reduced back to their ground state by accepting electrons from the redox electrolyte (usually an iodide/triiodide couple). The electrolyte itself is regenerated at the counter electrode by electrons arriving through the external circuit, completing the circuit and allowing continuous operation.

IV. Natural Dyes-A Sustainable Alternative

A. What Makes a Dye Photosensitive?

The dye photosensitizing performance of DSSCs depends on several interrelated photophysical and electrochemical properties. At its most basic, a dye should be able to strongly absorb visible light and transfer the excited electron to a semiconductor (most commonly titanium dioxide TiO_2) conduction band. That is, the dye molecules need to have a suitable electronic structure; in this case, the lowest unoccupied molecular orbital (LUMO) of the dye needs to be energetically located above the edge of the conduction band of TiO_2 to allow electron injection after photoexcitation. On the other hand, the highest occupied molecular orbital (HOMO) should be significantly below the redox potential of the electrolyte to ensure an effective regeneration of the dye and continuous charge transport [9].

DSSC photosensitizers and naturally occurring pigments typically possess anchoring groups like carboxyl, hydroxyl, or carbonyl. They enhance electronic coupling for rapid electron transfer and tether the dye to the TiO_2 surface to ensure stable adsorption. These characteristics are inherent in natural pigments such as anthocyanins, betalains, carotenoids, and chlorophylls. Anthocyanins, which are frequently isolated from fruit skins and flower petals, show potential as natural photosensitizers because of their abundance of hydroxyl and carbonyl groups and broad absorption in the visible spectrum. While carotenoids and xanthophylls, which are present in roots and fruits, provide complementary absorption and shield excited states from degradation, chlorophylls, which are found in green leaves, absorb mostly in the red and blue regions of the solar spectrum [12,13].

The stability issue is also significant; for a dye to produce a good quality, it cannot readily degrade when exposed to light or chemical exposure. The stability of natural dyes is poor in comparison with that of synthetic dyes or metal-organic dyes where it will easily become photobleached or desorbed from the TiO_2 substrate. Developments in co-sensitization (using mixtures of pigments), buffer layers, or hydrophobic modifications offset these shortcomings, enhancing the operational efficiency and durability. Buffer layers (e.g., alginate, spirulina) between TiO_2 and the dye prevent recombination losses and improve electron injection efficiency [12, 13].

B. Sources of Natural Dyes

DSSC construction-based dyes are available in all parts of the plant kingdom such as leaves, fruits, flowers, roots, and even some seeds. Diversity makes a sustainable, low-toxicity and easily available, local substitute for synthetic dyes. The major classes of natural photosensitizers are anthocyanins (flavonoids), betalains, carotenoids, and chlorophylls with spectral properties and their binding affinity to semiconductor materials [9, 12]. Some sources of natural dyes are discussed below:

- **Leaves:** Chlorophyll is omnipresent in plant leaves, and it's the green pigment that powers photosynthesis. Sources include spinach, grass, broccoli, and other leafy vegetables. The porphyrin structure of chlorophyll permits the efficient absorption of light, particularly toward the red and blue regions, and can provide optimal electron injection from photoexcitation [12, 13].
- **Fruits:** Anthocyanins, which impart blue, purple, and red coloration to fruit skins (e.g., berries, black grapes, mulberries, eggplant peels), are highly effective natural sensitizers. Betalains, abundant in beetroot and prickly pear, offer distinct red-violet hues and broad spectral coverage.

Fruit peels of avocado and pomegranate have also been tested, showing good efficiency due to favourable energy level alignment and pigment concentration [12, 13].

- **Flowers:** Many flower petals (hibiscus, saffron, blue pea, celosia, rosella, and mallow) are rich in anthocyanins and flavonoids. The molecular structures frequently contain carbonyl and hydroxyl groups, favouring stable binding to TiO_2 and robust light absorption across a wide spectral range. Yellow flowers (e.g., Canna lily, Cassia fistula) contain xanthophylls and other flavonoids, which complement anthocyanin absorption and enhance cell efficiency when co-sensitized [12, 13].
- **Roots and Tubers:** Carrots, sweet potato and turmeric root, contain carotenoids, provide orange to yellow pigments with long-wavelength absorption. Betalains which is found in beets are also widely researched for DSSC application [14].

Studies have often reported that combining multiple dye sources (e.g., green dyes from leafy vegetables with red/purple dyes from fruits or flowers) can synergistically widen absorption and improve solar conversion. Thus, a blend of 20% green dye (spinach) and 80% red dye (beetroot) was found to increase the performance of DSSC when compared to using a single dye [12].

C. Extraction and Preparation Methods

The extraction and preparation of natural dyes for DSSC is an engineered series which attempts to extract and utilize optimal quality photosensitizing pigments from botanical sources to obtain best photovoltaic performance. Studies commonly state that plant materials should be selected optimally for pigment-rich plants like leaves (chlorophylls), fruits and flowers (anthocyanins and betalains), and roots (carotenoids). These materials are washed to remove dust, soil, or surface contaminants and are then chopped into fine pieces to facilitate the release of pigments during subsequent extraction steps. Many (often specific) studies note that dark storage and gentle handling at this stage are key, as light and high temperatures can prematurely degrade sensitive pigments such as anthocyanins and chlorophylls [12, 14, 15].

The plant tissue is usually crushed or homogenised using either a mortar and pestle or a blender following an initial preparation. Such mechanical disruption ruptures structures of the cells, liberating the pigments and making them more accessible to solvents. The right extraction solvent is critical—ethanol and methanol are good for anthocyanins and chlorophylls, but more nonpolar solvents like acetone or hexane may also be employed for sources rich in carotenoids. In case of anthocyanins, it is a common practice to acidify the solvent with dilute hydrochloric or acetic acid, which stabilizes the pigment and sustains the vivid colour through the extraction [12, 14]. Extraction is done at room temperature or slightly elevated temperatures (up to 40°C), usually in the dark or low light conditions for several hours to overnight. This process safeguards sensitive pigments from photo-degradation and thermal decomposition. The mixture is stirred or allowed to soak statically, after which the extract is separated by filtration. Filter paper, centrifugation, or muslin cloth are employed to eliminate solid plant residues to obtain a coloured liquid extract of target dye molecules [12, 16].

For further concentration and purification, some experiments use evaporative techniques to remove excess solvent at low temperatures or liquid-liquid extraction and column chromatography to segregate the wanted pigments from the impurities and degradation products. pH adjustment is available where necessary, since many natural dyes, particularly anthocyanins, work best in acidic conditions. The stability of the final dye solution is critical, so it is kept in dark, airtight containers, typically refrigerated until use. UV-Vis spectroscopic examination of the dye is done then to find ideal absorption peaks and to ensure pigment stability [12, 14].

The prepared dye solution is subsequently used to sensitize the TiO_2-coated photoanode of the DSSC. The substrate is immersed in the dye solution for a soak period ranging from several hours up to a full day, ensuring strong adsorption of the pigment molecules onto the nanoporous TiO_2 surface. Upon adsorption, the electrode is rinsed (typically using the same solvent) to remove any loosely bound dye and is dried before full assembly of the device. Other studies report the use of mixed dye sources (e.g. beetroot and turmeric) and adjusting extraction pH to maximize efficiency, with acidic conditions yielding a marked effect on the power conversion. Based on several studies, this general protocol is developed to ensure reproducibility while achieving high dye activity for DSSC effective performance [12, 14, 16].

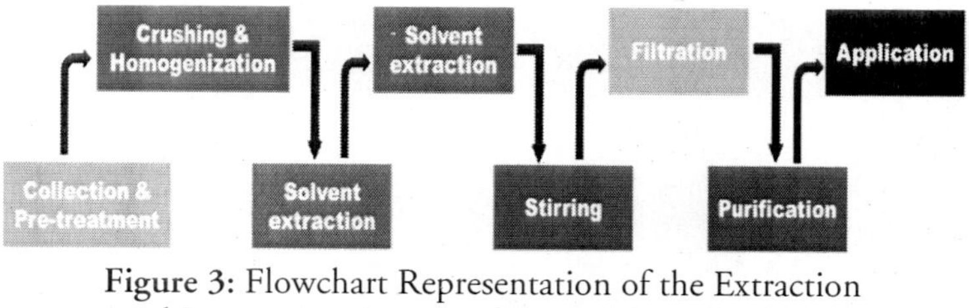

Figure 3: Flowchart Representation of the Extraction and Preparation Process of Natural Dyes for DSSC

V. Performance and Efficiency Factors

A. Light Absorption and Electron Transfer

Visible light is absorbed by the dye, not by the wide band-gap semiconductor. The dye injects electrons into the conduction band of the semiconductor, enabling visible-light solar energy conversion [17, 18].

The dye (D), usually a ruthenium complex or organic dye, is adsorbed on the surface of a nanocrystalline TiO_2 semiconductor.

When sunlight strikes the cell, the dye absorbs a photon (hν) and becomes excited:

$$c + h\nu \rightarrow c^{*}$$

Where D is ground-state dye and D* is excited dye molecule.

The excited dye molecule injects an electron into the conduction band of TiO_2:

$$c^{*} \rightarrow c^{+} + e^{-}(\text{s “n}_{2})$$

The electron enters the conduction band of TiO_2 and the dye becomes oxidized (D^+). The injected electron travels through the TiO_2 network to the transparent conducting electrode, flows through the **external circuit**, and reaches the counter electrode. At the counter electrode, electrons reduce the redox electrolyte (iodide/triiodide system):

$$h_3^{-} + 2e^{-} \rightarrow 3h^{-}$$

B. Stability and Degradation of Natural Dyes

The stability of dye sensitized solar cell using natural dyes is lower than the synthetic one. The stability of natural dyes such as anthocyanins, chlorophyll, curcumin, and betalains depends on factors such as light, pH, solvent polarity and temperature [19-21]. For example, carotenoids are highly sensitive to photodegradation, while tannins and anthraquinones are relatively stable. The dyes often undergo photobleaching, desorption from TiO_2 surfaces, and oxidative degradation, thermal decomposition etc which leads to decrease in cell efficiency [22].

The stability of natural dyes in DSSCs can be improved by

- Adding functional group to dye molecules for strong binding with TiO_2 layer.
- Using dyes with protective nanoparticles to reduce degradation.

- Adding stabilizers
- Balancing the pH of the surrounding condition.

The cell stability of natural dye-based DSSC can be improved by employing a UV filter and proper sealing. The instability of DSSCs has been organized into two main groups in review: physical and chemical instabilities. Physical instabilities include the possible loss of I_3^- ion from the electrolyte solution at the ambient temperature, exposure to UV radiations, and humidity, all which can permanently damage the cell. For chemical instability, decomposition of dyes, electrolyte and counter electrodes, and the effect of oxygen, water, and temperature during the operation, etc., have been discussed in detail [23].

C. Improving Efficiency Through Material Choices

There are many parameters that has role on the efficiency of the performance of the cell, as has been studied in various research literatures over the time, like light source temperature, frequency, humidity, open circuit voltage, short circuit current and thermal stresses on working factors of DSSC such as fill factor and efficiency [24].

To improve the effectiveness of DSSCs, the oxidised dye must be quickly reduced to its ground state after electron injection to prevent recombination. The redox mediator potential should be close to the dye's ground state to get the maximum photovoltage. The working electrode should be uniformly sensitized to prevent the trapping of TiO_2 nanoparticles. The performance and efficiency of the electrode can be improved by using the various materials such as nanotubes, nanowires of carbon, and graphene using gel electrolyte and quasi-solid electrolytes instead of a liquid one, such as gel electrolyte and quasi-solid electrolytes. Scandium doped TiO_2 can increase the efficiency. FeN/N-doped graphene counter electrode and Counter electrode doped with $NiCO_2S_4$ can have maximum efficiency [25].

Several materials can be used to increase the efficiency of dye-sensitized solar cells (DSSCs):

(a) **Materials for Photoanodes: (i) TiO_2:** Because of its stability and large surface area, it is widely used (ii) ZnO: It is another well-liked choice with a high electron mobility (iii) SnO_2: It provides excellent stability and electron mobility.

(b) **Materials for Sensitizers:** (i) Dyes based on Ruthenium: They are very effective, but costly (ii) Metal-free organic dyes: These are economical substitutes for ruthenium (iii) Natural dyes: less effective but better for the environment.

(c) **Materials for electrolyte:** Iodide/triiodide (I-/I3-) is a frequently used redox couple as electrolyte material. Cobalt-based electrolytes are effective and promising substitute, while solid-state electrolytes provide increased efficiency and stability.

(d) **Materials for Counter Electrodes:** (i) Platinum (Pt): It is expensive but has high catalytic activity (ii) Carbon-based materials: They are low-cost substitutes with strong catalytic activity (iii). Transition metal compounds: They provide excellent stability and efficiency.

(e) **Extra Resources:** (i) Graphene: It improves conductivity and electron mobility (ii). Nanotubes: They enhance electron collection and transport (iii) Doping materials: They improve the stability and conductivity of the material. To stop deterioration, the materials should be compatible. To guarantee long-term performance, selected materials should have high stability. They should be cost-effective. The efficiency of DSSCs can be greatly increased through careful material selection and optimisation, increasing their viability for a range of applications.

VI. Environmental and Economic Benefits

DSSCs are helpful for both the environment and the economy. They use clean energy from the sun and are made from simple, low-cost materials. This makes them good for nature and affordable for people. They help reduce pollution and also save money. Below are some points that explain their environmental and economic benefits.

DSSCs can convert sunlight into electricity using simple, low-cost, and eco-friendly materials. Unlike traditional silicon solar cells, which are expensive and complex to manufacture, DSSCs can be made using cheaper and widely available components such as natural dyes, TiO_2, and simple electrolytes [26, 27]. These natural dyes act like chlorophyll in plants—they absorb sunlight and help convert it into electrical energy. Many DSSC materials are biodegradable and non-toxic, which makes them environmentally friendly [28]. They can also be produced in various colors and on flexible substrates, making them suitable for windows, walls, and portable devices. Because of their low production cost, simple fabrication, and use of safe materials, DSSCs are seen as a promising and sustainable alternative for future solar technologies [26, 29].

DSSCs help reduce the carbon footprint because they generate electricity from sunlight without burning fossil fuels. This means they do not release harmful greenhouse gases such as carbon dioxide [30]. They also require less energy to manufacture compared to silicon-based cells, resulting in a shorter energy payback time [30]. Since many DSSC materials are natural, low-cost, and partly biodegradable, their overall environmental impact is lower. Thus, DSSCs produce clean energy and help keep the planet healthy [30, 31].

DSSCs have promising applications in developing regions and education. In areas with limited or unreliable electricity, DSSCs provide an affordable and accessible source of clean energy [32]. They can be assembled using local materials and simple tools, empowering communities to build and maintain their own solar devices. In education, DSSCs are widely used in school and university projects to demonstrate renewable energy concepts, nanotechnology, and sustainability [33]. Through such applications, DSSCs not only generate power but also promote environmental awareness and scientific learning.

VII. Challenges and Future Prospects

Although DSSCs have many advantages, they also face challenges. Researchers are working to improve their efficiency, stability, and long-term durability. With new materials and designs, DSSCs have a bright future in clean energy. A major challenge is their limited lifespan and weather resistance. The dye and electrolyte components can degrade when exposed to heat, light, or humidity, leading to lower efficiency over time [34, 35]. Liquid electrolytes may leak or evaporate, while temperature fluctuations can further reduce stability [36]. Therefore, DSSCs are currently less durable than silicon solar cells, but new research on solid and polymer electrolytes aims to overcome these issues [37].

Scaling up DSSCs from lab-sized prototypes to large modules introduces challenges in maintaining efficiency and uniformity. Large-area DSSCs often suffer from reduced performance due to recombination losses, sealing issues, and non-uniform dye coverage [38, 39]. Furthermore, cost-effective large-scale production remains a challenge due to the precision required in coating and assembly [40]. Future research focuses on developing more stable and efficient dyes, solid-state electrolytes, and hybrid structures that combine DSSCs with other photovoltaic technologies such as perovskite solar cells [41]. These hybrid systems can improve light absorption and efficiency while enhancing durability. With these advances, DSSCs could become more powerful, affordable, and sustainable for widespread use [41].

VIII. Conclusion

In summary, natural dye-sensitized solar cells (NDSSCs) represent a promising step toward sustainable and environmentally friendly photovoltaic technology. Although their current power conversion efficiencies remain lower than those of synthetic dye systems, advances in dye extraction, molecular modification, and device engineering continue to improve their performance. The use of naturally derived pigments such as anthocyanins, chlorophylls, carotenoids, and betalains not only reduces toxicity and environmental impact but also promotes the development of renewable materials in energy applications. With continued research focused on optimizing dye adsorption, charge transfer dynamics, and long-term stability, NDSSCs have the potential to evolve into practical, low-cost alternatives for future green energy solutions.

Acknowledgement

Authors acknowledge Students Project Scheme 2024-2025 of B. Borooah College (Autonomous) for providing financial support during this study.

References

[1] K.K. Jaiswal, C.R. Chowdhury, D. Yadav, R. Verma, S. Dutta, K.S. Jaiswal, N. Sangmesh B.K. and S. Karuppasamy, Renewable and sustainable clean energy development and impact on social, economic, and environmental health., *Energy Nexus*, 7, 100118 (2022).

[2] A. Mdallal, A. Yasin, M. Mahmoud, M.A. Abdelkareem, A.H. Alami and A.G. Olabi, A comprehensive review on Solar Photovoltaics: Navigating generational shifts, innovations and sustainability, Sustainable Horizons, 13, 100137 (2025).

[3] A. Blakers, N. Zin, K.R. McIntosh, and K. Fong, High Efficiency Silicon Solar Cells, Energy Procedia, 33, 1 (2013).

[4] L.C. Andreani, A. Bozzola, P. Kowalczewski, M. Liscidini and L. Redorici, Silicon solar cells: toward the efficiency limits, *Advances in Physics X*, 4, 1548305 (2018).

[5] A. Kumar, M. Bieri, T. Reindl and A.G. Aberle, Economic viability analysis of silicon solar cell manufacturing: Al-BSF versus PERC, *Energy Procedia,* 130, 43 (2017).

[6] M. Tawalbeh, A. Al-Othman, F. Kafiah, E. Abdelsalam, F. Almomani and M. Alkasrawi, Environmental impacts of solar photovoltaic systems: A critical review of recent progress and future outlook, *Science of the Total Environment,* 759, 143528 (2020).

[7] S. Al-Ali, A.G. Olabi and M. Mahmoud, A review of solar photovoltaic technologies: developments, challenges, and future perspectives, *Energy Conversion and Management X,* 101057 (2025).

[8] K. Sharma, V. Sharma and S.S. Sharma, Fabrication of CdS quantum dots: structural and optical properties, *Nanoscale Research Letters*, 13, (2018).

[9] S. Rahman, A. Haleem, M. Siddiq, M.K. Hussain, S. Qamar, S. Hameed and M. Waris, Research on dye sensitized solar cells: recent advancement toward the various constituents of dye sensitized solar cells for efficiency enhancement and future prospects, *RSC Advances,* 13, 19508 (2023).

[10] W.A. Ayalew and D.W. Ayele, Dye-sensitized solar cells using natural dye as light-harvesting materials extracted from Acanthus sennii chiovenda flower and Euphorbia cotinifolia leaf, *Journal of Science: Advanced Materials and Devices,* 1, 488-494 (2016).

[11] G. Zacharias, M.B. Hansen, *et al.*, The Adsorption of Ru-Based Dyes on the TiO_2 Surface to Enhance the Photovoltaic Efficiency of Dye-Sensitized Solar Cell Devices, Molecules, vol. 30, no. 6, 1312, (2024).

[12] Shukor, N.I.A., Chan, K.-Y., Thien, G.S.H., Yeoh, M.-E., Low, P.-L., Devaraj, N.K., Ng, Z.-N. and Yap, B.K. (2023). A Green Approach to Natural Dyes in Dye-Sensitized Solar Cells. Sensors, 23(20), 8412.

[13] U. Mahajan, K. Prajapat, M. Dhonde, K. Sahu and P.M. Shirage, Natural dyes for dye-sensitized solar cells (DSSCs): An overview of extraction, characterization and performance, Nano-Structures & Nano-Objects, 37, 101111 (2024).

[14] C. Cari, N. Khairuddin, T.Y. Septiawan, P.M. Suciatmoko, D. Kurniawan, and A. Supriyanto, The preparation of natural dye for dye-sensitized solar cell (DSSC), AIP Conference Proceedings 2014 (1), 020106.

[15] I. W. Estiningtyas, N. Kusumawati, P. Setiarso, S. Muslim, N.T. Rahayu, R.N. Safitri, N. Zakiyah and F.F. Fachrirakarsie, Effect of natural dye combination and pH extraction on the performance of dye-sensitized photovoltaics solar cell, *International Journal of Renewable Energy Development*, 12, 1054 (2023).

[16] Zhenzhen Yang, Shanmin Gao, Tao Li, Fa-Qian Liu, Yang Ren, and Tao Xu, Enhanced Electron Extraction from Template-Free 3D Nanoparticulate Transparent Conducting Oxide (TCO) Electrodes for Dye-Sensitized Solar Cells, *ACS Applied Materials & Interfaces* 2012 4 (8), 4419-4427.

[17] D. Zheng, X. Yang, L. Čuček, J. Wang, T. Ma and C. Yin, Revolutionizing dye-sensitized solar cells with nanomaterials for enhanced photoelectric performance, *Journal of Cleaner Production*, 464, (2024), 142717.

[18] S.N. Karthick, K.V. Hemalatha, S.K. Balasingam, and M.C. Franklin, Dye-Sensitized Solar Cells: History, Components, Configuration, and Working Principle, Interfacial Engineering in Functional Materials for Dye-Sensitized Solar Cells, Wiley, (2019) https://doi.org/10.1002/9781119557401.ch1.

[19] K. Prajapat, U. Mahajan, A. Kumar, M. Dhonde, K. Sahu, S. Vyas, Y.M. Riyad, and Z.M. El-Bahy, Next-generation counter electrodes for dye-sensitized solar cells: A comprehensive overview, *Sustainable Materials and Technologies*, 42, 2024, e01178.

[20] C. Zhong, W.B. Hu and Y.F. Cheng, On the essential role of current density in electrocatalytic activity of the electrodeposited platinum for oxidation of ammonia, *Journal of Power Sources,* 196, 8064–8072 (2011).

[21] Fahmid Kabir, Serajum Manir, Md. Mossaraf Hossain Bhuiyan, Sikandar Aftab, Hamidreza Ghanbari, Amirhossein Hasani, Mirette Fawzy, G.L. Thushani De Silva, Mohammad Reza Mohammadzadeh, Ribwar Ahmadi, Amin Abnavi, Abdelrahman M. Askar and Michael M. Adachi, Instability of dye-sensitized solar cells using natural dyes and approaches to improving stability–An overview, *Sustainable Energy Technologies and Assessments*, 52, Part C, 2022, 102196.

[22] S. A. Abrol, C. Bhargava and P.K. Sharma, Material and its selection attributes for improved DSSC, *Materials Today: Proceedings,* 42, 1477-1484 (2021).

[23] P. Singh and A. Sachdeva, DSSC Fundamentals and Optimization Materials: A Review, IOP Conference Series: Earth and Environmental Science, vol. 1285, no. 1, p. 012003, 2024,

[24] K.J. Hwang, J.H. Lee, and H.K. Kim, "What is necessary to fill the technological gap to design sustainable dye-sensitized solar cells?" *Sustainable Energy & Fuels,* 7, 1456–1471 (2023).

[25] M.I. Hossain, *et al.*, A comprehensive review on recent progress in dye-sensitized solar cells, *Chemical Society Reviews,* 50, 12318–12367 (2021).

[26] A.F. Nogueira and J.N. Freitas, Eco-friendly materials for DSSC fabrication, *Journal of Photochemistry and Photobiology A: Chemistry*, 383, 112026 (2019).

[27] Indian Express, IIT Hyderabad develops low-cost, eco-friendly solar cells using kumkum dye (2019). Available: https://indianexpress.com/article/technology/science/iith-develops-low-cost-eco-friendly-solar-cells-using-kumkum-dye-5864501

[28] X. Zhou, *et al.*, Life cycle assessment of dye-sensitized solar cells: Environmental performance and carbon footprint, *Renewable Energy,* 162, 1463–1473 (2020).

[29] J.Y. Lim, *et al.,* Environmental performance of window-integrated DSSC modules in Malaysia, *ASEAN Energy Journal,* 5(4), 233–241 (2022).

[30] J.N. Freitas, *et al.,* Sustainable DSSCs: From materials to devices, *Renewable and Sustainable Energy Reviews,* 156, 111982 (2022).

[31] M.M. Rahman, *et al.*, Potential of dye-sensitized solar cells for rural electrification in developing countries, *Energy Reports*, 7, 2119–2128 (2021).

[32] Y. Hu and T. Zhang, Using dye-sensitized solar cells in STEM education to teach renewable energy concepts, *Journal of Chemical Education,* 96(5), 1043–1049 (2019).

[33] F. Bella, *et al.,* Solid-state electrolytes for long-lasting DSSCs, *ChemSusChem,* 8(11), 3668–3681 (2015).

[34] Q. Zhang, *et al.,* A review on long-term stability of dye-sensitized solar cells, *Nano-Micro Letters,* 10(3), 51 (2018).

[35] K. Sharma, *et al.*, Degradation and stability of DSSCs under environmental stress, *Solar Energy Materials and Solar Cells,* 209, 110420 (2020).

[36] J. Li, *et al.*, Gel and polymer electrolytes for durable DSSCs, *Electrochimica Acta,* 452, 142304 (2023).

[37] E.M. Barea, *et al.*, Challenges of scaling up dye-sensitized solar cells, *Nanoscale Research Letters,* 13, 247 (2018).

[38] T.V. Nguyen, *et al.*, Performance decline of large-area DSSCs: Causes and mitigation, *Solar Energy,* 260, 84–93 (2023).

[39] B. O'Regan and M. Gratzel, Scaling dye-sensitized solar cells: Industrial and manufacturing perspectives, *Progress in Photovoltaics,* 27(8), 637–650 (2019).

[40] R. Giannuzzi, *et al.*, Hybrid perovskite–DSSC devices for enhanced stability and performance, *Advanced Energy Materials,* 12(6), 2103341 (2022).

[41] S. Mathew, *et al.*, Next-generation dye-sensitized solar cells: Materials and hybrid architectures, *Chemical Reviews,* 121(5), 3473–3550 (2021).

CHAPTER 6

Sustainable Materials for Solar Panels: From Science to Society

Dhrubajyoti Choudhury
Department of Physics, Baosi Banikanta Kakati College, Assam, India
E-mail: dhrubajyotichoudhury90@gmail.com

ABSTRACT: The worldwide shift toward renewable power has established solar photovoltaics (PV) as the primary technology which fights climate change while lowering carbon pollution and protecting natural resources from depletion. The advancement of materials science has enabled scientists to create better solar panels through their work on material development. The chapter reviews all scientific progress made in solar panel materials starting from traditional crystalline silicon and thin-film technologies up to current perovskite and organic photovoltaic and tandem solar cell developments. The discussion examines how new materials solve three essential problems which include improving energy conversion efficiency, reducing production expenses and solving environmental issues. The review demonstrates how scientific discoveries meet public needs to create a sustainable future through solar panel materials. The review in this chapter demonstrates that PV material innovations serve two purposes by advancing energy science and creating solutions for worldwide environmental problems including climate change and resource depletion and environmental damage. The research examines both laboratory-based material selection and their resulting social effects in the community. The paper identifies three essential elements for a sustainable clean energy transition which include obtaining raw materials through sustainable methods and toxic element reduction and recycling technology advancement. The review examines how material science interacts with policy frameworks to demonstrate that successful implementation requires unified approaches which link new technologies to public approval and business viability and environmental sustainability.

Keywords: Photovoltaics Materials, Thin-Film Technologies, Perovskite Solar Cells, Tandem Solar Cells, Solar Panel Materials, Crystalline Silicon, Band Gap.

I. INTRODUCTION

The energy landscape across the world is changing drastically that urgent need to mitigate climate change, reduce greenhouse gas emissions and shift away from finite fossil fuel resources. Among renewable energies, solar photovoltaics (PV) has turned up as a basis solution because of its capacity to be changed in size, abundance and rapidly decreasing costs [1]. Photovoltaic devices convert sunlight into electrical energy (electricity) where photons generate electron-hole pairs. The generating charge carriers are separating with an internal electric field which driving an electric current through an external circuit. The stability, efficiency and affordability of PV devices are dictated primarily by the choice of materials. The semiconductor absorber must exhibit an optimal band gap for solar spectrum absorption, chemical stability and long carrier diffusion lengths. In addition, equally important to transparent conductive oxides (TCOs), metallic contacts and encapsulation layers, which together define the long-term performance of solar modules.

This chapter provides a detailed academic and research-oriented discussion of solar PV materials, which covering both established and emerging technologies. It presents the fundamental physics of energy conversion, explores conventional crystalline silicon and thin-film technologies. Also analyses the extraordinary rise of perovskite and organic photovoltaics and examines tandem architectures that exceed single-junction efficiency limits. Quantitative models examine sustainability challenges including raw material scarcity, recyclability and toxicity. Finally, the interplay between materials innovation and real-world frameworks is discussed and underscoring the importance of policy and fair access to technology.

II. Conventional Materials in Photovoltaics

A. Crystalline Silicon Solar Cells

The crystalline silicon (c-Si) has dominated PV markets for decades accounting for more than 90% of installed capacity worldwide [2]. Silicon is non-toxic, abundant and benefits from a mature industrial supply chain, largely developed by the semiconductor industry.

Silicon solar cells exist in two primary forms:

Monocrystalline Silicon (mono-Si): High-purity wafers derived from the Czochralski process. They offer higher efficiencies (~24–26%) due to lower defect densities.

Multicrystalline Silicon (mc-Si): The material is cast into ingots which are then cut into sliced wafers. They are less expensive but typically yield lower efficiencies (~20–22%) because of grain boundary defects.

The theoretical efficiency of silicon solar cells is limited by the Shockley-Queisser (SQ) limit, which defines the maximum conversion efficiency of a single-junction solar cell under standard illumination as ~33% [3]. This limit arises because photons with energy below the band-gap are not absorbed. Also, those with excess energy above the band-gap are lost as heat.

The power conversion efficiency (PCE) can be expressed as:

$$\eta = \frac{P_{out}}{P_{in}} = \frac{qV_{oc}J_{sc}FF}{P_{sun}}$$

Where, q is the elementary charge, V_{oc} is the open-circuit voltage, J_{sc} is the short-circuit current density, FF is the fill factor, and P_{sun} is the incident solar irradiance (~1000 W/m^2 under AM1.5G conditions).

Commercial silicon solar cells have approached 26.8% efficiency in laboratory conditions [4], close to the SQ limit. Ongoing research focuses on passivated emitter rear contact (PERC) cells, hetero junction with intrinsic thin layer (HIT) cells and silicon-based tandem structures.

B. Thin-Film Technologies

To minimize the material usage and the cost of processing, thin-film solar cells came as the alternatives of silicon. The thin-film absorber is deposited in a few microns thickness as a layer in contrast to thick silicon wafers.

- **Cadmium Telluride (CdTe):** The most commercially successful thin-film technology is CdTe solar cells which have efficiencies of over 22% [5]. The direct band Gap of CdTe is 1.45 eV which is close to that of the ideal solar cell that can be made using a single junction. Nevertheless, the toxicity of cadmium and the lack of tellurium is an environmental and economic issue.
- **Copper Indium Gallium Selenide (CIGS):** CIGS cells utilize quaternary chalcopyrite alloys, and band-gaps in the range of 1.0 to 1.7 eV can be obtained based upon the In/Ga ratio [6]. Record efficiencies are over 23%, however, large-scale implementation is limited by the limited availability of indium and gallium and complexities of deposition of the thin-film.
- **Amorphous Silicon (a-Si:H):** Amorphous silicon is done using lithe hydrogenated films which are deposited through plasma-enhanced chemical vapor deposition (PECVD). Although cheap and easily scalable, it has, among other things, relatively low efficiencies (around 10–12%), and the Staebler-Wronski effect, causing light-induced degradation.

III. EMERGING MATERIALS IN PHOTOVOLTAICS

The limitations of silicon and traditional thin-film absorbers have push forward intensive research into new classes of semiconductors including perovskites, organic photovoltaics and tandem configurations.

A. Perovskite Solar Cells

Hybrid organic-inorganic perovskites (represented by the general formula ABX_3) have revolutionized PV research. Here, A is an organic cation such as methylammonium ($CH_3NH_3^+$), B is typically Pb^{2+} and X is a halide ion such as I^- or Br^-.

Within just over a decade, perovskite solar cells (PSCs) have achieved laboratory efficiencies of 26.1%, rivalling silicon [7]. Their exceptional optoelectronic properties include:

- High absorption coefficients ($>10^5$ cm^{-1})
- Long carrier diffusion lengths (exceeding 1 μm)
- Tunable bandgaps (1.2–2.3 eV)
- Low-temperature solution fabrication

The current density in PSCs can be modelled as:

$$J = \int_0^\infty \Phi(E)A(E)\eta_{sep}(E)\eta_{coll}(E)dE$$

Where, $\Phi(E)$ is the solar photon flux, $A(E)$ is the absorption probability, η_{sep} is the exciton separation efficiency, and η_{coll} is the carrier collection efficiency.

Here the challenges include environmental instability (sensitivity to moisture, heat and oxygen) and the toxicity of lead. Tin-based perovskites (Sn^{2+}) and all-inorganic $CsPbI_3$ variants are being explored as alternatives.

B. Organic Photovoltaics (OPVs)

The organic photovoltaics rely on semiconducting polymers or small molecules as absorbers. It is observed that recent developments have pushed efficiencies beyond 19% [8]. Their advantages include mechanical flexibility, low-cost solution processing and the potential for semi-transparent modules. OPVs are excitonic devices meaning photons create bound excitons that require dissociation at donor-acceptor interfaces. Bulk heterojunction architectures maximize exciton separation and charge collection. However,

OPVs face challenges including limited operational lifetimes and narrower absorption spectra compared to inorganic semiconductors.

C. Tandem Architectures

The Shockley-Queisser limit can be exceed by tandem (multi-junction) cells which combine multiple absorbers with complementary band gaps.

- **Perovskite-Silicon Tandems:** Efficiencies above 33% have been achieved [9].
- **All-Perovskite Tandems:** Two perovskites with different band-gaps enable lightweight and high-efficiency modules.
- **Multi-Junctions:** Laboratory efficiencies exceed 40% but costs are prohibitive for terrestrial deployment.

The theoretical efficiency of a two-junction tandem is:

$$\eta_{tandem} = \frac{q(V_{oc,1}J_{sc,1} + V_{oc,2}J_{sc,2})}{P_{sun}}$$

where subscripts 1 and 2 indicate the top and bottom cells.

IV. Sustainability Challenges in Material Selection

Although the quest to achieve greater efficiency has led to remarkable advancements with regard to the growth of photovoltaic technologies, such a pursuit usually comes with the challenge of high sustainability that cannot be ignored. Of great concern is the toxicity of some elements. In CdTe modules in particular cadmium and perovskite solar cells in particular lead, which increases the environmental and health risks unless they are well handled in the entire lifecycle of the panels. The shortage of such critical raw materials like indium, gallium and tellurium which are important in CIGS and CdTe technologies is another established issue. But small in world supply, posed threats of bottlenecks in supply-chains, and long-term economical restraints. Besides these material restrictions, solar panels recyclability has been a serious issue. Even though silicon-based modules have begun to enjoy the advantages which are being introduced by developing recycling protocols, thin-film and perovskite gadgets do not have efficient programs. Re-use of commercial viability of recycling routes that give rise to anxieties of impending electronic refuse build-up. The LCA studies have shown that photovoltaic systems produce much less carbon dioxide in the order of 20–40 g CO_2-equivalent per kWh in comparison with coal which is over 1000 g CO_2-equivalent per kWh. Nevertheless, the total ecological footprint of solar technology can be reduced only when the processing, source and end-of-life of materials is optimally structured. Thus, sustainable materials research must not only aim at improving the performance of the devices but also at toxicity, scarcity and recyclability so that the photovoltaics is indeed the long-term solution to the resource depletion and the climate change.

V. Societal and Policy Dimensions

Since it influences how innovations in material science are translated into real-world impact, therefore the societal and policy dimensions surrounding PV deployment are equally important. One of the key issues lies in the geopolitical concentration of supply chains. A small number of nations dominate the production of silicon wafers. Critical raw minerals like tellurium, cobalt, and indium are mainly exploited in areas with weak governments [11]. This concentration poses ethical questions about work practices and communication difficulties in addition to making people vulnerable to price volatility and trade disruptions. Therefore, policy actions are essential to resolving these issues. PV adoption will be accelerated and investment in clean energy infrastructure will be encouraged with the use of tools including carbon pricing schemes, feed-in tariffs, renewable portfolio standards, and subsidies. Furthermore, the transition to a circular economy framework for resource security and waste reduction is becoming essential. Where end-of-life PV modules are methodically collected, repurposed and reintegrated into production. Public acceptance of solar technology is also shaped by social factors. Including the visibility of large solar farms, land-use considerations and the perception of long-term sustainability. Therefore, effective integration of technological innovation, policy mechanisms and societal engagement will determine the future of solar materials. The extent to which solar photovoltaics can deliver on their promise of providing clean and equitable energy for all.

VI. Future Directions

PV research must address both performance and sustainability in an integrated manner ensuring that technological upgradation push into long-term benefits for society in future. One critical area is the development of lead-free perovskites such as tin- and germanium-based compounds. That aims to retain the exceptional optoelectronic properties of lead halide perovskites while eliminating concerns of toxicity and environmental hazards. In addition to this, significant progress is being made in tandem architectures where perovskite-silicon and all-perovskite combinations are used to exceed the Shockley-Queisser efficiency limit and achieve record-breaking power conversion efficiencies. The scalable deposition techniques, including inkjet printing, blade coating and slot-die coating are equally important which allow the transition of laboratory-scale devices into industrially manufacturable modules. Keeping also in mind that reducing material wastage and overall costs. During minimization of waste production at the end of operational lifespan of a solar photovoltaic module, to recover critical raw materials such as indium, silver and gallium are become essential by the adoption of closed-loop recycling processes. The integration of photovoltaics with energy storage systems is expected to play a vital role in enabling continuous 24/7 renewable energy supply. Addressing the intermittency of solar generation and encouraging more constructive energy infrastructures. In a nutshell, these directions highlight the dual pathway of increasing efficiency and ensuring sustainability which will define the future generation of solar technologies.

VII. Conclusion

PV technology has the ability to supply clean, plentiful, and equitable energy for future generations by linking science and society. Alignment with generation demands, sustainable supply chains, and reliable recycling technologies are ultimately necessary for a sustainable solar future. This chapter has provided a comprehensive academic analysis of materials for solar PV, highlighting the interplay between fundamental physics, materials chemistry and engineering. The debate was grounded in scientific principles by the presentation of equations like the Shockley-Queisser limit and photocurrent models. Conventional crystalline silicon and thin-film absorbers continue to dominate. However, perovskites, OPVs and tandem solar cell represent breakthrough innovations.

References

[1] M.A. Green, Solar Cells: Operating Principles, *Technology, and System Applications,* Prentice Hall, 1982.

[2] Fraunhofer ISE, Photovoltaics Report, 2023.

[3] Shockley, W. and Queisser, H.J., Detailed Balance Limit of Efficiency of p-n Junction Solar Cells, *J. Appl. Phys.,* 32, 510 (1961).

[4] NREL, Best Research-Cell Efficiency Chart, *National Renewable Energy Laboratory,* 2023.

[5] M. Woodhouse, *et al.*, Supply-Chain Dynamics of CdTe Photovoltaics, *Prog. Photovolt. Res. Appl.*, 21, 12 (2013).

[6] P. Jackson, *et al.*, Properties of Cu(In, Ga)Se_2 Solar Cells with New Record Efficiencies up to 23.35%, *Prog. Photovolt*, 25, 12 (2017).

[7] NREL, Perovskite Solar Cell Research, 2023.

[8] G. Li *et al.*, 19% Efficiency Organic Solar Cells Enabled by an Alloy Acceptor, Joule 5, 15 (2021).

[9] A. Al-Ashouri, *et al.*, Monolithic Perovskite/Silicon Tandem Solar Cell with >29% Efficiency by Enhanced Hole Extraction, *Science*, 370, 1300 (2020).

[10] D. Nugent and B. Sovacool, Assessing the Lifecycle Greenhouse Gas Emissions from Solar PV and Wind Energy, *Energy Policy*, 52, 557 (2013).

[11] International Energy Agency (IEA), *Solar PV Global Supply Chains Report*, 2022.

CHAPTER 7

Advanced Carbonaceous Materials for Next Generation Aluminium-Sulfur Battery

P. Priyanka*, B. Kamum and S.K. Das
Department of Physics, Tezpur University, Tezpur, Assam, India
*E-mail: pdmnprnk18@gmail.com

ABSTRACT: Excessive use of fossil fuels has led to environmental pollution as well as their depletion. To overcome this crisis, there is a need to develop competent energy storage technologies. The commercialised lithium-ion batteries that are used today is limited by their high cost and safety concerns, and hence fail to fulfil the growing demands of the society. In the recent years, metal-sulfur batteries have emerged as a promising technology for the next generation high energy density batteries with better safety, lower cost and longer cycling stability. Amongst them, aluminium-sulfur battery has the advantage of aluminium being highly abundant, thus resulting in cost effectiveness. Moreover, both aluminium and sulfur have very high theoretical energy density. However, such batteries face certain difficulties relating to the high reactivity of aluminium, sluggish kinetics of sulfur, capacity fading and also the dissolution of polysulfides. Thus, there lies a requirement for cathode materials to be engineered as sulfur hosts which can entrap sulfur and improve the efficiency, stability, along with the lifespan of the battery. Herein, a focus on the carbon-based materials as potential cathodes in aluminium- sulfur batteries is discussed. Carbon-based materials have the capacity of holding sulfur and thus act as host for sulfur. They have the properties of higher surface area and large pore volume which enable them to confine sulfur and its reaction products. The higher electronic conductivity of carbon hosts also increases the rate of reaction. Thus, this chapter aims to highlight the various carbon-based cathode materials that can be used for high performance aluminium-sulfur batteries.

Keywords: Metal-sulfur Batteries, Aluminium-sulfur, Polysulfides.

I. INTRODUCTION

Energy is a basic requirement for both industrial and economic development. Today, most of our energy needs are being fulfilled by fossil fuels. However, the continuous dependency on these fossil fuels has several issues which have raised major concerns owing to pollution of the environment, scarcity of resources and energy security. The global energy policies have also laid emphasis on reducing the carbon dioxide emissions which is mainly caused by these energy sources. Also, by 2030, the world's energy demand is predicted to be doubled [1]. Hence, renewable energy sources like wind and solar energy are being explored in order to meet the ever-increasing demand of energy. In order to overcome all these challenges, it becomes crucial to be able to store ample amounts of energy from the renewable sources. Energy storage systems serve as a link which connects energy supply with demand. Common examples of energy storage technologies include electric vehicles and portable electronics like smartphones and laptops.

In energy storage systems, energy can be stored in the form of chemical, thermal, mechanical or electrical storage systems. Among these, the chemical storage systems mainly include the batteries that are one of the most promising technologies for electrochemical storage due to their use in a vast range of fields from small portable devices to large scale applications [2, 3]. Electrochemical systems enable reversible redox reactions between the electrode and electrolyte. Such a system mainly consists of three components: the anode, the cathode and the electrolyte that conducts ions. Electrochemical energy systems are broadly classified as batteries, fuel cells and supercapacitors. The electrochemical storage systems have their roots which date back to about 200 years [4]. Modern battery technology is marked by the pioneering work of Alessandro Volta. Battery technologies are classified as primary and secondary (rechargeable) batteries. The rechargeable batteries allow the chemical reactions to be reversible unlike the primary batteries and hence are preferred. Following this, many rechargeable battery technologies have been developed to ensure improvements in performance, cost effectiveness, environmental benefits and durability.

Among the rechargeable batteries, lithium-ion batteries (Li-ion) are extensively used today and for this invention, John B. Goodenough, M. Stanley Whittingham and Akira Yoshino was awarded the Nobel Prize in Chemistry in 2019 [5]. Since its commercialization in 1990 by Sony, the performance of Li-ion batteries has been greatly enhanced through modifications of electrode and electrolyte [2]. Such batteries have the advantages of having high energy density, increased cycle life and high energy efficiency. However, in order to improve the energy density of such batteries and meet the growing demand for energy, there lies the need to develop cathode materials with the characteristics of having higher capacity and energy density along with being cost effective.

Sulfur is one such material which is emerging as an attractive cathode material. It has a high theoretical capacity of 1675 $mAhg^{-1}$, is a low-cost material with high abundancy and has both safety and environmental benefits [6]. Also, the lithium-sulfur (Li-S) batteries can achieve a theoretical specific energy of 2600 $Whkg^{-1}$ and a working voltage of 2.1 V [7]. In a lithium-sulfur battery, lithium is used as the anode, sulfur is used as cathode and an organic electrolyte is used which facilitates the movement of lithium ions across the electrodes. A 2-stage discharge process takes place in such cells which is denoted as follows [8]:

$$S_8 \rightarrow Li_2S_6 \rightarrow Li_2S_4$$
$$Li_2S_4 \rightarrow Li_2S_2/LiS$$

During the charging process, all the reactions are reversed.

However, such batteries face many challenges like low conductivity as both sulfur and the final product (Li_2S_2/LiS) are insulators, polysulfide shuttle effect which arises due to the solubility of polysulfides in the electrolyte, disruption in structural integrity of the cathode, dendrite formations at the anode that lead to the degradation of the lithium anode, compromising with the battery quality as well as safety [9]. Thus, it can be concluded that although, the Li-ion batteries are widely used today, they are expensive, have resource availability problems and safety and flammability issues. In order to overcome these challenges, researchers are moving towards systems that are cheaper and based on highly abundant materials on earth. One such system is the aluminium-based batteries, where aluminium is the most abundant metal on the earth's crust, is inexpensive and have higher safety. Also, each aluminium ion has 3 electrons (Al^{3+}) which allows for a very high theoretical and volumetric capacities of 2980 $mAhg^{-1}$ and 8046 $mAhcm^{-3}$ respectively which in turn accounts for batteries with high energy densities [10].

A lot of research activities are currently undergoing on aluminium-ion (Al-ion) batteries with various types of cathode materials like metal oxides, metal sulfides, carbon-based materials and so on. However, such insertion-type cathodes pose challenges in such batteries. Due to the high charge density of aluminium ions and the large size of the ionic aluminium species, intercalation reactions are difficult to occur. In order to overcome this, a conversion-type cathode is required such as sulfur which has certain advantages as mentioned above. As we know that the Li-S batteries have also been explored, but coupling aluminium with sulfur provides significant advantages regarding cost and safety as well. Such a battery system has the potential to offer high energy density, is cost effective and provides a safer energy storage system [11].

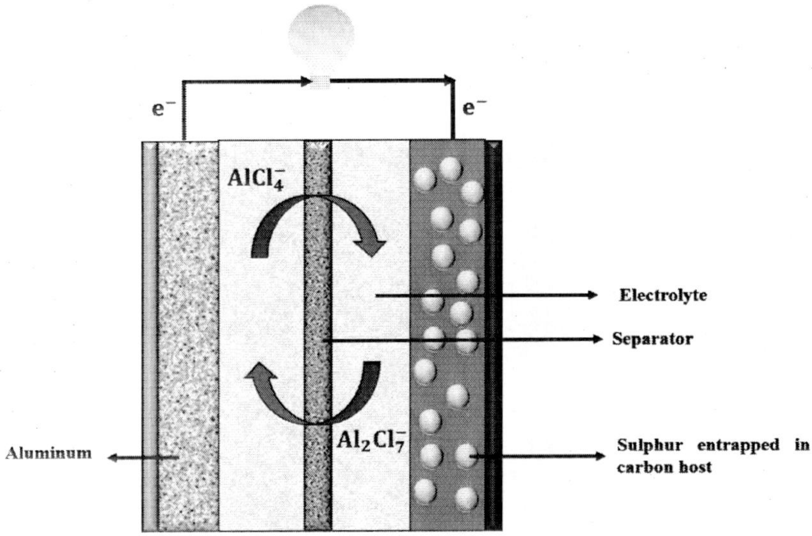

Figure 1: An Al-S battery system

Yet, despite its numerous advantages, the aluminium-sulfur (Al-S) system faces challenges due to several factors like the low conductivity and sluggish kinetics of sulfur, structural degradation, shuttle effect due to polysulfide dissolution and so on [11]. All these factors pave the way towards engineering new cathode materials which can act as hosts for sulfur in order to be able to address all the aforementioned issues and provide superior electrochemical performance. This is where the carbon structures come into play. Different types of carbon structures are utilized to function as sulfur hosts in order to be able to mitigate the issues and enhance the performance of Al-S battery systems [12]. This chapter focusses on the various carbon hosts that can be used for entrapping sulfur and thus contribute towards effective and efficient energy storage technologies.

II. Why Carbon Hosts?

Carbon materials have higher surface areas and electronic conductivity which makes them attractive host to be able to entrap aluminium polysulfides and thus prevent polysulfide dissolution in Al-S batteries [12], which is a major concern in such batteries. In addition to this, the carbonaceous materials have a structure that is highly tuneable. This allows researchers to be able to modify the various interactions that will help in increasing the performance of the battery system. The carbon hosts also have high thermal stability along with having high resistance to damage under mechanical stress, which enables them to provide structural integrity and operation across all range of temperatures. Hence, in order to be able to bring out the best of such Al-S battery technologies, various carbon materials have been explored, which will be discussed in the next section.

III. Types of Carbon Hosts for Sulfur Cathode

The emergence of the Al-S batteries began during the 1980s and since then, researchers have been putting immense efforts in improving the design of Al-S batteries by various optimizations [13]. Hence, a number of carbon-based structures can be explored to serve as sulfur hosts.

Graphene oxide and reduced graphene oxide (GO/rGO) play a promising role as sulfur host for Al-S batteries in enhancing cycle stability, electrochemical performance and significant capacity retention. The porous structure of graphene, along with the incorporation of a polar functional group in GO/rGO, facilitates the confinement or adsorption of polysulfides, inhibiting their dissolution [14]. Furthermore, to attain enhanced cycle stability, Zheng *et al.* fabricated a carbon-based composite cathode incorporating elemental sulfur, graphene and cobalt sulfide along with the reduced graphene oxide (rGO) coated separator forming a sandwich structure which significantly inhibits polysulfide diffusion and lessens the battery's polarization. Consequently, the Al–S battery has an impressive discharge capacity range of 676–1145 $mAhg^{-1}$ at 50 mAg^{-1} with a steady cycle life of 38 cycles [15].

The carbon nanotubes including the single walled carbon nanotubes (SWCNTs) also can be used as sulfur hosts for Al-S batteries. The porous structure of single walled carbon nanotubes physically limit the sulfur. Additionally, the SWCNTs facilitate conductive paths for electron transport, resolving the concerns involving low electronic conductivity and the dissolution (shuttle effect) of polysulfides [16]. In this context, Yu *et al.* developed rechargeable room temperature Al-S batteries with a meticulously developed matrix which was prepared by ditributing sulfur/Al[EMI]Cl_4 slurry in carbon nanofibre (CNF). S/CNF (sulfur in carbon nanofibre) cathode was used in order to enhance the material's surface area and structures to mitigate the repercussions of polysulfide dissolution. Furthermore, he used a SWNT layer on the separator to prevent dispersed polysulfides from migrating to ensure Al-S battery cycling. Reportedly having a high initial discharge capacity, the Al-S cell's discharge capacity drastically drops to significantly lower than 300 $mAhg^{-1}$ [17].

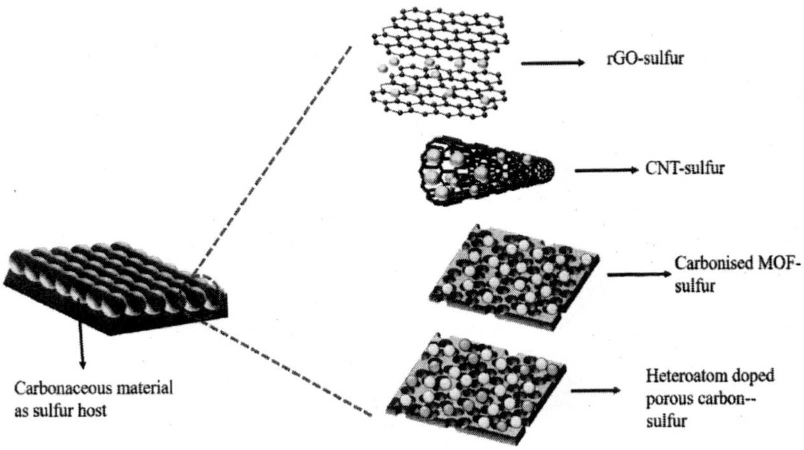

Figure 2: Different Types of Carbonaceous Materials Used as Sulfur Host

Another type of carbon host that can be used are the carbonized metal-organic frameworks (MOF). Carbonization of MOF results in the formation of a porous carbon structure that holds sulfur, while residual metal nanoparticles from the MOF facilitate catalytic activity and enhance conductivity, which contributes to better cycle stability and capacity [18]. Furthermore, to demonstrate that the carbon materials' pores serve as an effective inhibitor to thwart the polysulfide shuttle effect, Guo *et al.* reported the fabrication of copper decorated carbonized-MOF (HKUST-1-C) as sulfur host in Al-S batteries. The initial discharge capacity of S@HKUST-1-C reached 1200 $mAhg^{-1}$ and it maintained the discharge capacity of 600 $mAhg^{-1}$ at the 75th cycle and 460 $mAhg^{-1}$ at the 500th cycle, even at 1 Ag^{-1} current density. Additionally, Cu, which is present in the HKUST-1, enhances electrochemical activity and optimizes the reversible nature of sulfur by creating a polysulfide based ionic clump [19].

Doped carbon matrices are essential in aluminum-sulfur (Al-S) and analogous batteries (such as lithium-sulfur and sodium-ion) for increasing their cycle stability by mitigating the intrinsic negative aspects of sulfur cathodes, particularly their low electrical conductivity and the "shuttle effect" of intermediate polysulfides. Numerous studies reported an analogous conclusion, such as research conducted by Wang *et al.*, who synthesized nitrogen-doped 3D porous carbon material as an efficient sulfur host for high-performance Al-S batteries [20]. Furthermore, among the configurations of N dopant in carbon matrices, pyrrolic N and pyridinic N are synergistic as they primarily provide active sites for electrochemical reactions and aid in mitigating the shuttle effect, whereas graphitic N primarily maintains the electrical conductivity and the structural stability/integrity of the carbon matrices providing a good volume buffering for prolonged cycle stability of batteries [21, 22]. Heteroatom doped-porous carbon based supports when used as carbon hosts for the sulfur cathode, can lead to significant improvements in both the batteries' lifespan and their ability to retain specific capacity over time. In this context, Lv *et al.* have reported the fabrication of porous Co, N-doped graphene-carbon nanotubes (CoN-GC) hybrid as a sulfur host in Al-S batteries. The initial specific capacity of Al-S batteries with S@CoN-GC reached 1330 $mAhg^{-1}$ at a current density of 300 mAg^{-1}. Subsequently, it retained the specific capacity of 278 $mAhg^{-1}$ despite performing 2000 cycles [23]. These substantially prolonged lifespan of Al-S batteries were due to the introduction of CoN_4 catalytic active sites to the CoN-GC hybrid, the rates of polysulfide conversion escalated as well as rapid electron transfer and aluminum ion diffusion were enhanced. Approaching towards the green based derived sulfur host, Faheem *et al.* reported the fabrication of N, S-doped, ordered porous carbon (DSC-800) made from thiourea-modified silk cotton as a highly efficient sulfur host in Al-S batteries. The initial discharge capacity of DSC-800-S were found to be 535 $mAhg^{-1}$ at a current density of 100 mAg^{-1}. Moreover, the DSC-800-S maintained its discharge capacity of 200 $mAhg^{-1}$ after 160 cycles [21]. The high surface area and heteroatom-enriched carbon matrix in DSC-800 enhanced sulfur entrapment and electron transfer, allowing for a good cycle life and improved capacity retention. These research studies yield an invaluable perspective towards the various ways that carbon-based cathode support is utilized in Al-S batteries.

The above mentioned carbonaceous materials of different dimensions: 1D, 2D and 3D can be imposed as various architectures for posing as hosts for holding sulfur atoms, that can help in enhancing the overall electrochemical performance in Al-S batteries. The 1-D structures such as carbon nanotubes have high electrical conductivity along with continuous pathways for electrons to transport. Also, the morphology of carbon nanotubes enhances the confinement of sulfur and improves the electron transport. However, they do have the disadvantage of having limited polysulfide confinement and hence limits cycling stability. On the other hand, the 2-D structures such as graphene, graphene oxide and reduced graphene oxide have a layered structure and high surface area with a large number of polar functional groups. This permits stronger adsorption of polysulfides along with suppressing of the shuttle effect which is a common problem encountered in aluminium- sulfur battery systems. The 3-D carbon structures which includes the carbonized metal-organic frameworks (MOF) and heteroatom doped-porous carbon structures can accomodate higher sulfur content due to their porous structure along with structural stability and faster ion transport. When combined with 1D or 2 D structures, they can be considered as the most efficient sulfur host in order to achieve superior electrochemical performance in aluminium-sulfur batteries.

IV. Conclusion

In summary, the chapter illustrates about various carbon materials that can be used for hosting sulfur. It can be concluded that carbon based materials play a crucial role as a cathode material (sulfur) host by entangling thc aluminum polysulfides due to their large surface area and good electronic conductivity. A number of studies have shown that such an approach has been demonstrated to be potentially effective in inhibiting the dissolution of polysulfides [25]. Numerous research studies have been performed utilizing a diverse array of carbon materials, including graphene oxide and reduced graphene oxide (GO/rGO), carbon nanotubes including both SWCNTs and MWCNTs, carbon doped with heteroatoms, etc. which have been utilized as sulfur hosts in aluminum-sulfur batteries [26, 27]. Given their distinct mechanical, electrical, and morphology properties, these carbon materials functions as diffusion inhibitors for polysulfides. These beneficial aspects of various carbon frameworks have demonstrated enormous potential in optimizing the specific capacity, cycle stability and energy density of Al-S batteries.

Acknowledgement

Padmini Priyanka acknowledges DST Inspire Fellowship, IF240146.

References

[1] A. Moradzadeh, M. Nazari-Heris, B. Mohammadi-Ivatloo and S. Asadi, Energy storage fundamentals and components, *Energy storage in energy markets,* 23 (2021).

[2] S.T. Revankar, Chemical energy storage, *Storage and Hybridization of Nuclear Energy,* 177, (2019).

[3] J.B. Goodenough, Electrochemical energy storage in a sustainable modern society, *Energy & Environmental Science,* 7, 14 (2014).

[4] O. Haas and E.J. Cairns, Electrochemical energy storage, *Annual Reports Section "C" (Physical Chemistry),* 95, 163 (1999).

[5] J.H. Kim, Grand challenges and opportunities in batteries and electrochemistry, *Frontiers in Batteries and Electrochemistry,* 1, 1066276 (2022).

[6] A. Manthiram, Y. Fu and Y.S. Su, Challenges and prospects of lithium–sulfur batteries, *Accounts of chemical research,* 46, 1125 (2013).

[7] A. Benítez, J. Amaro-Gahete, Y.C. Chien, Á. Caballero, J. Morales and D. Brandell, Recent advances in lithium-sulfur batteries using biomass-derived carbons as sulfur host, *Renewable and Sustainable Energy Reviews,* 154, 111783 (2022).

[8] Manthiram, A., Fu, Y., Chung, S.H., Zu, C. and Su, Y.S., Rechargeable lithium–sulfur batteries, *Chemical reviews,* 114, 11751 (2014).

[9] S. Yari, A. Conde Reis, Q. Pang and M. Safari, Performance benchmarking and analysis of lithium-sulfur batteries for next-generation cell design, *Nature Communications,* 16, 5473 (2025).

[10] Y.H. Wu, W.J. Lin, T.H. Tsai and M.C. Lin, Unveiling the reaction mechanism of aluminum and its alloy anode in aqueous aluminum cell, *ACS Applied Energy Materials,* 7, 3957 (2024).

[11] X. Yu, M.J. Boyer, G.S. Hwang and A. Manthiram, Room-temperature aluminum-sulfur batteries with a lithium-ion-mediated ionic liquid electrolyte, Chem, 4, 586 (2018).

[12] D. Zhang, X. Zhang, B. Wang, S. He, S. Liu, M. Tang and H. Yu, Highly reversible aluminium–sulfur batteries obtained through effective sulfur confinement with hierarchical porous carbon, *Journal of Materials Chemistry A,* 9, 8966 (2021).

[13] X. Zhang, Y. Tong, J. An, F. Cheng, Z. Wu, Y. Xue, Z. Huang, Z. Fang and S. Jiao, Foundations, Design Strategies, and Further Considerations for High-Energy Al-S Batteries, *Electrochemical Energy Reviews,* 8, 15 (2025).

[14] A.I. Kamisan, T.I.T. Kudin, A.S. Kamisan, A.F.C. Omar, M.F.M. Taib, Hassan, O.H., ... and M.Z.A. Yahya, Recent advances on graphene-based materials as cathode materials in lithium-sulfur batteries, *International Journal of Hydrogen Energy,* 47, 8630 (2022).

[15] X. Zheng, R. Tang, Y. Zhang, L. Ma, X. Wang, Y. Dong, ... and L. Wei, Design of a composite cathode and a graphene coated separator for a stable room-temperature aluminum–sulfur battery, *Sustainable Energy & Fuels,* 4, 1630 (2020).

[16] T. Patodia, K.B. Sharma,, S. Dixit, S. Katyayan, G. Agarwal, A. Jain, ... and B. Tripathi, Carbon nanotube-sulfur nanocomposite electrodes for high energy–foldable lithium sulfur battery, *Materials Today: Proceedings,* 42, 1638 (2021).

[17] X. Yu and A. Manthiram, Electrochemical energy storage with a reversible nonaqueous room-temperature aluminum–sulfur chemistry, *Advanced Energy Materials,* 7, 1700561 (2017).

[18] Z. Li, C. Li, X. Ge, J. Ma, Z. Zhang, Q. Li, ... and L. Yin, Reduced graphene oxide wrapped MOFs-derived cobalt-doped porous carbon polyhedrons as

sulfur immobilizers as cathodes for high performance lithium sulfur batteries, *Nano Energy*, 23, 15 (2016).

[19] Y. Guo, H. Jin, Z. Qi, Z. Hu, H. Ji and L.J. Wan, Carbonized-MOF as a sulfur host for aluminum–sulfur batteries with enhanced capacity and cycling life, *Advanced functional materials*, 29, 1807676 (2019).

[20] J. Wang, J. Xu, Z. Huang and G. Fan (2021). Preparation of nitrogen-doped three-dimensional hierarchical porous carbon/sulfur composite cathodes for high-performance aluminum-sulfur batteries. Fullerenes, *Nanotubes and Carbon Nanostructures*, 29(1), 39–45.

[21] M. Faheem, A. Hussain, A.S.R. Hakim, M. Ali, A.S. Hakeem and M.A. Aziz, Thiourea-modified silk cotton carbon as an efficient sulfur host: A green approach for advanced aluminum-sulfur battery, *Biomass and Bioenergy*, 203, 108329 (2025).

[22] W. Dong, Z. Wu, X. Zhu, D. Shen, M. Zhao, F. Yang, ... and S. Yang (2024). Synergistic Pyridinic N/Pyrrolic N configurations in rGO/CNT composite sulfur hosts for high-performance lithium-sulfur batteries. *Chemical Engineering Journal*, 488, 150872.

[23] J. Lv, G. Lai, T. Yang, X. Sun, F. Liu, W. Wu, ... and N. Chen, Enabling long cycle aluminum-sulfur batteries via structurally stable Co, N-doped graphene-CNTs covalently bonded hybrid, *Chemical Engineering Journal*, 494, 152811 (2024).

[24] S. Debnath, M. Horscheck-Diaz, D.J. Searles and M. Hankel (2021). Carbon nitrides as cathode materials for aluminium ion batteries. *Carbon*, 183, 546-559.

[25] N. Deka and G.K. Dutta, Recent advancement in carbon-based materials as sulfur hosts for rechargeable aluminum–sulfur (Al–S) batteries, *Current Opinion in Electrochemistry*, 38, 101222 (2023).

[26] Y. Ai, S.C. Wu, F. Zhang, X. Zhang, R. Li, Y. Lan, ... and W. Wang, Bifunctional TiN@ N-doped-graphene catalyst based high sulfur content cathode for reversible Aluminum-Sulfur batteries, *Energy Storage Materials*, 48, 297 (2022).

[27] H. Li, J. Lampkin and N. Garcia-Araez, Facilitating Charge Reactions in Al-S Batteries with Redox Mediators, *ChemSusChem*, 14, 3139 (2021).

CHAPTER 8

Cr-doped CeO_2 Ceramics: Promising Materials for Next-Generation Dielectric Capacitors and Energy Storage Applications

R. Basumatary and R. Brahma*

Department of Physics, Bodoland University, Kokrajhar, Assam, India

*E-mail: rajibbasumatary201@gmail.com; rajibhcu@gmail.com

ABSTRACT: The next generation of energy storage devices is gradually focusing on solid-state fuel systems due to their high charge density, fast charging capability, and slow discharging characteristics. In this work, Cr-doped CeO_2 ceramics were synthesized via the solid-state route to evaluate their charge storage potential. X-ray diffraction (XRD) analysis revealed that the $Ce_{1-x}Cr_xO_2$ (0% and 18%) ceramics exhibit a cubic fluorite phase structure with phase space group Fm-3m. The large crystallite size is estimated for Cr-doped CeO_2, approximately 76 nm, along with a microstrain of 10.6×10^{-3}. Field Emission Scanning Electron Microscopy (FESEM) revealed a fine microstructure with an average grain size of about 0.25 μm in the Cr-doped CeO_2 ceramics. Dielectric measurements showed that Cr doping enhances the dielectric properties while maintaining a low dielectric loss. The AC conductivity of both the pristine and doped samples indicates a hopping mechanism associated with short-range charge transport. Furthermore, the electric-field-dependent polarisation response is improved with the Cr incorporation, yielding a maximum recoverable energy density of 0.074 mJ/cm² at an applied field of 5 kV/cm and an efficiency (η) of ~64.91%. The overall improvement in polarisation, energy density and efficiency highlights the potential of Cr-doped CeO_2 ceramics as promising candidates for advanced dielectric capacitors and energy storage devices in the next generation electronic applications.

Keywords: Cr-doped CeO_2 Ceramics, Dielectric Properties, AC Conductivity, P-E Response, Energy Storage Devices.

I. INTRODUCTION

In modern times, the use of small electronic devices has increased significantly in our daily lives, as well as across various industrial and educational sectors. To keep these devices operational without any disruption, a continuous supply of energy is essential, which is primarily provided by batteries. However, conventional energy storage systems mainly rely on chemical reactions, posing potential risks such as explosions under high temperatures, along with limitations in renewability and efficiency. As the world moves toward renewable energy and widespread electrification, the demand for advanced battery technologies is skyrocketing to ensure sustainable energy solutions. To meet these demands, researchers are exploring innovative energy storage systems that can operate efficiently in every condition of environments, offering fast charging, slow discharging, and high charge density. Among these emerging technologies, solid-state batteries have gained significant attention. By utilizing a solid electrolyte instead of a liquid one, they promise enhanced energy density, improved safety, and greater overall performance, making them a potential breakthrough for the future of energy storage. Within this context, CeO_2 and doped ceria variants have been widely studied. When an appropriate amount of transition metals is doped, a significant vacancy is created in the CeO_2 materials [1]. Apart from that, multi-doped ceria exhibits significantly enhanced ionic conductivity for intermediate-temperature operation. However, CeO_2's mixed ionic-electronic behaviour, high oxygen-storage/release capability, and surface reactivity make it a useful interlayer, interface stabiliser, catalyst support, or coating to suppress deleterious reactions at electrode-electrolyte boundaries and to improve cycle life in composite or coated electrodes [2]. However, ionic conductivity at room temperature of pure ceria is usually low, requiring doping where Ce^{4+} is reduced to Ce^{3+} ions, producing electronic conduction that degrades electrolyte selectivity and cell efficiency [3]. Ceria-based ceramics can also suffer from interface resistance with lithium metal or other electrodes, mechanical brittleness, and sintering challenges [4-5]. The literature shows CeO_2-based materials are promising as electrolytes, interlayers, and functional additives in Solid-State Batteries (SSB) architectures, but practical deployment for room-temperature Li-metal solid electrolytes requires careful doping and interface design to avoid electronic leakage, maintain chemical stability, and achieve high ionic conductivity [6-7]. Despite the increase in conductivity with the increase in ionic/electronic conductivity, the leakage current also increases, which is the major challenge [8-9]. Ghosh *et al.* [10] reported that Cr ions improved the energy storage capabilities of CeO_2 materials. Harish *et al.* [11] noted that the lower dielectric loss of the as Cr ions increases in CeO_2 materials. Pedroli *et al.* [12] reported that the leakage current is directly related to the dielectric loss.

Although extensive research has been conducted on Cr-doped CeO_2, its ferroelectric behaviour and potential for energy storage applications remain relatively unexplored. This present work mainly focuses on the role of Cr ion incorporation in CeO_2-based materials, highlighting its structural and electrical behaviour, with particular relevance to capacitor dielectric and advanced energy storage technologies.

II. Experimental Technique

The polycrystalline $Ce_{1-x}Cr_xO_2$ (X = 0% and 18%) ceramics were synthesised via the solid-state route. The starting materials Chromium Oxide (Cr_2O_3, 99%) and Cerium carbonate ($Ce_2(CO_3)_3 \cdot xH_2O$, 99.9%) were purchased from Sigma-Aldrich (Merck). The synthesis process involved mixing and grinding to achieve the desired stoichiometry ratio. To maintain the homogeneity, a few drops of the organic solvent acetone were added during the mixing process. The mixing was then ground for 8 hours, and during the grinding, the mixture, the acetone had evaporated over time. To eliminate organic solvent and carbonate impurities, the samples were pre-calcined at 300°C and 500°C for 3 hours and 5 hours, respectively. Afterwards, the materials were subjected to final calcination at 900°C for 8 hours. Then the fine powder was collected and mixed with PVA, and the mixture was pressed into cylindrical disc pellets of dimensions of 1 mm in thickness with 10 mm diameter, and then sintered at 1100°C for 2 hours. This calcination process helps further enhance the material's properties and structural integrity.

III. Characterizations

The Cr-doped CeO_2 compound's phase compositions and structure are characterised at room temperature utilizing a M/s Rigaku, TTRAX-III 18 kW, Japanese X-ray powder diffractometer. The scanning range is set to 20° to 80° with a step size of 0.02°, and Cu-Kα radiation with a wavelength of 1.5406 Å is used. The surface morphology of the sample is examined via a Field Emission Scanning Electron Microscopy (FE-SEM) (M/s Zeiss Sigma, Germany). The frequency-dependent dielectric studies were performed at room temperature across an applied frequency range of 100 Hz to 1 MHz using the LCR meter (M/s Nihon Denkei Co., Ltd., HIOKI IM3536, Japan). The ferroelectric properties of the samples are characterised by measuring the polarisation vs. electric field hysteresis loop with a ferroelectric tester (M/s Radiant Technologies, Inc., Precision LCII, USA).

IV. Results & Discussion

A. Powder X-ray Diffraction Analysis

The X-ray diffraction patterns of $Ce_{1-x}Cr_xO_2$ (X = 0% and 18%) compounds are illustrated in Figure 1(a). The peaks distinctly show the cubic fluorite structure of the CeO_2 compounds, even in the Cr-doped sample no additional peaks were found. The recorded peaks are consistent with the JCPDS No. 81–0792. The magnified view of peak (111) in Figure 1(b) reveals a slight shift towards a higher value of 2θ, indicating the lattice construction of the CeO_2 matrix [13]. This may be due to the formation of Ce^{3+} ions in the process of incorporation of Cr^{3+} ions in the CeO_2 matrix. The incorporation of Cr^{3+} ions tends to replace Ce^{4+} ions in the matrix because Cr^{3+} ions have a smaller ionic radius compared to Ce^{3+} ions [14].

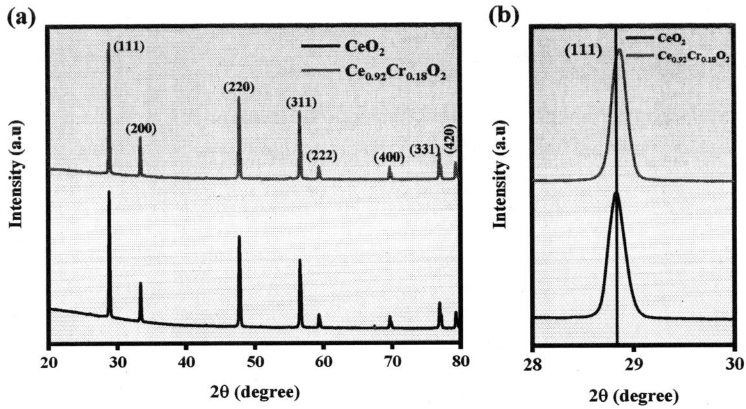

Figure 1: (a) Cubic Fluorite X-ray Patterns of $Ce_{1-x}Cr_xO_2$ (X = 0% and 18%) Compounds (b) Magnified View of (111) Peak

The average grain size and microstrain of the compounds are estimated by utilizing the Williamson-Hall method as shown in Figure 2(a) [15]. It is observed that the average crystallite size was found to increase in the Cr-doped CeO_2 compound. This might be due to the formation of oxygen-related

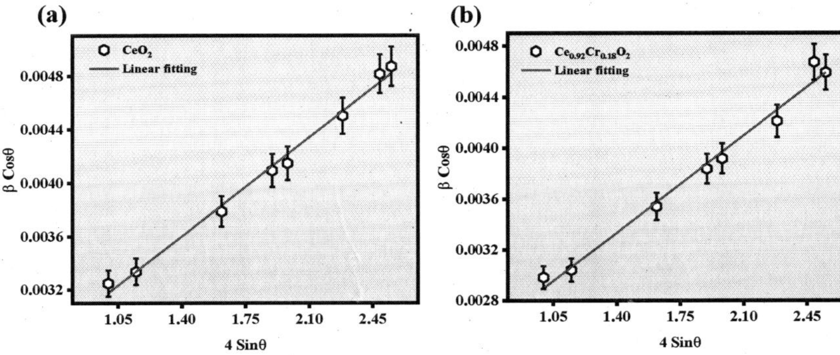

Figure 2(a-b): Williamson-Hall Plot of $Ce_{1-x}Cr_xO_2$ (X = 0% and 18%) Compounds

vacancies in the matrix, where, during the crystallization of the compound, the Cr^{3+} ions replace the Ce^{4+} in the matrix and in order to maintain charge neutrality, the vacancies are created. The microstrain of the Cr-doped CeO_2 is also slightly high, indicating the lattice distortion due to the Cr ion incorporation. The estimated parameters are summarised in Table 1.

Table 1: Calculated Parameters of XRD Peaks of Cr-doped CeO_2 (0% and 18%) Compounds

Samples	*CeO_2*	*$Ce_{0.82}Cr_{0.18}O_2$*
Av. Crystallite Size (nm)	65	76
Strain ($\times 10^{-3}$)	0.00106	1.09

B. Structural Analysis

Figure 3(a) reveals the particle shape and size of the $Ce_{1-x}Cr_xO_2$ (18%) composite; the shape of the grains ranges from spherical to irregularly shaped, with a large number of porous between the grains. The lognormal distribution of the grains in Figure 3(b) reveals that most of the grains lie in the 0.20–0.25 micrometre range.

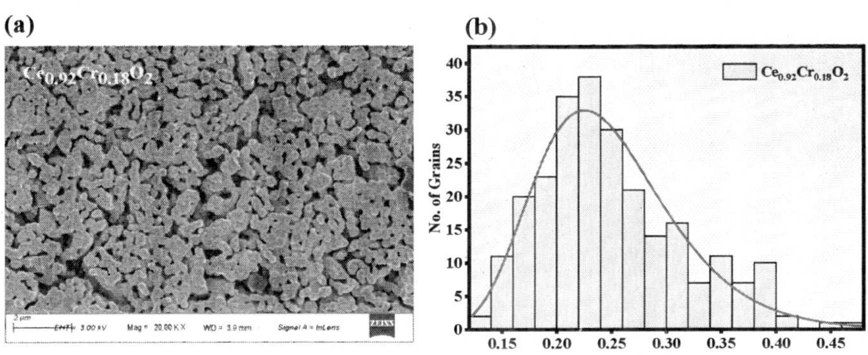

Figure 3: (a) FE-SEM Image and (b) Log Normal Distribution of Grain Size

C. Dielectric Properties

Figure 4(a) shows the dielectric constant at room temperature of $Ce_{1-x}Cr_xO_2$ compounds. The dielectric constant at lower frequencies is high in both samples, and at the Cr ions-doped CeO_2 compound, the dielectric constant further increases up to 110. At higher frequencies, the dielectric constant became independent of frequency. This is the typical dielectric spectrum of the ceramics. The dielectric behaviour of the ceramic can be explained by the

Maxwell-Wagner model, where the ceramic consists of two layers, a conductive and a resistive (non-conductive) layer [16]. To verify the Maxwell-Wagner model of the prepared samples, the Havriliak-Negami model is applied to fit the dielectric spectra. The model is well fitted with the dielectric constant spectra. The calculated parameters of the dielectric constant are listed in Table 2. Sharma *et al.* [17] noted the enhancement of the dielectric constant as a certain amount of Cr^{3+} ions were doped. The Cr^{3+} may increase the space charge in the host materials as defects created within the matrix. The formation of defects in CeO_2 materials by Cr doping is reported by Chen *et al.* in their paper [18].

Figure 4(b) illustrates the dielectric loss of the ceramic $Ce_{1-x}Cr_xO_2$ (X = 0% and 18%) compound. The spectra revealed a unique type of energy loss (negative loss) in the ceramics at lower frequencies. Such spectra were recorded by Axelrod [19] at the interval of temperature of –100°C to 300°C, where the porosity of the sample was attributed to the local non-compensated matrix-anchored charges, which accumulated both inside the

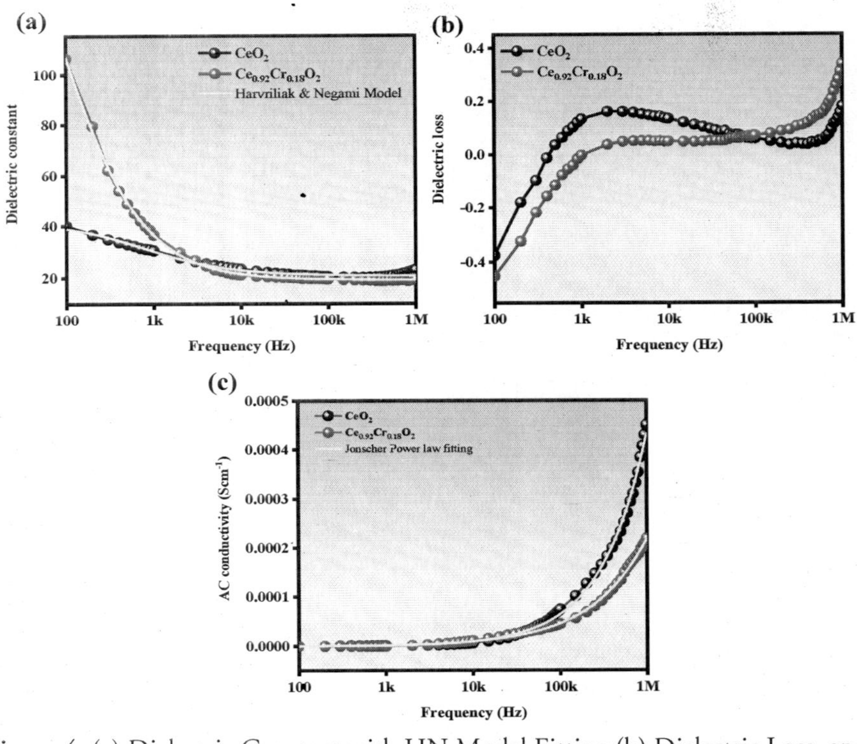

Figure 4: (a) Dielectric Constant with HN Model Fitting (b) Dielectric Loss, and (c) AC Conductivity with Jonscher's Power Law Fitting of $Ce_{1-x}Cr_xO_2$ Compounds

porous materials and on the interface of the pores. Therefore, according to Bartkowska *et al.* [20], the intrinsic polarization within the material aligns opposite to the applied electric field. Herein, the dielectric loss is negative at lower frequencies, and as the frequency increases, the loss traces to the positive side. This trend may be due to the space charge of the compounds, which couldn't line up with the applied field, leads to the space charge polarization oppose the external field. The low dielectric loss at room temperature at lower frequency indicates the low leakage current in the lower frequency region [12]. At higher frequencies, there is a rise in dielectric loss, indicating an increase in leakage current due to the creation of Ce^{3+} ions.

Figure 4(c) displays the frequency-dependent ac conductivity measured at room temperature of samples CeO_2 and Cr-doped CeO_2 compounds. In the lower frequency region, the conductivity is almost constant and gradually increases towards higher frequencies as electron hopping frequency increases [21]. This type of conductivity may be attributed to bound charge carriers trapped in oxygen defects, where the trapped electrons begin hopping after passing the threshold frequency. The ac conductivity at higher frequencies can be explained by Jonscher's power low as expressed in Equation 3.

$$\sigma_{AC} = \sigma_{DC} + A\omega^{s} \quad \text{... (3)}$$

Here, $\omega = 2\pi f$ is the angular frequency of the applied electric field, f refers to the linear frequency, A represents constant parameters, and s is the fitting parameter value ranging between $0 < s < 1$. Where s value is defined as the conduction mechanism at higher frequency, s value of both the samples is less than 1, which signifies the short-range-hopping mechanism. Figure 4(c) revealed that the Cr-doped CeO_2 sample reduces ac conductivity at higher frequencies, indicating the hampering of the electrical conductivity at higher frequencies. The reduction of the ac conductivity may be due to the blockage of hopping charge carriers, attributed to the large defects that segregate at the grain boundaries [21].

Table 2: Comprehensive Dielectric Parameters of Pristine CeO_2 and Cr-doped CeO_2 Compounds

Parameters	*CeO_2*	*$Ce_{0.82}Cr_{0.18}O_2$*
Dielectric relaxation strength ($\Delta\varepsilon$)	20.48	188.11
Relaxation time (τ)	0.00185	0.01354
Width of loss peak (α)	0.746	0.894
AC conductivity fitting value (s)	0.679	0.826

D. Ferroelectric Properties

Figure 5(a-b) illustrates the field-dependent polarization of pristine CeO_2 and the Cr-doped CeO_2 compound at room temperature. Ferroelectric measurements are conducted at a frequency of 10 Hz, with variations of applied voltage. The measurements are recorded within the limits of breakdown voltage. The maximum polarization value increased at Cr doped sample, this may be the Cr ions distort the CeO_2 matrix while incorporation, which leads to the increase on non-centrosymmetric. Moreover, the Cr ions suppressed the leakage current which blocks the hopping of Ce^{4+}/Ce^{3+} ions in the CeO_2 sample [22]. Such improved P-E loops observed in the 18% at Cr ion-doped samples. This signifies that the Cr ions help reduce the leakage current in the CeO_2 compound. To assess the energy storage capability within the samples, we employed the calculation of recoverable energy storage density. By applying the same procedure as previously published [1]. The

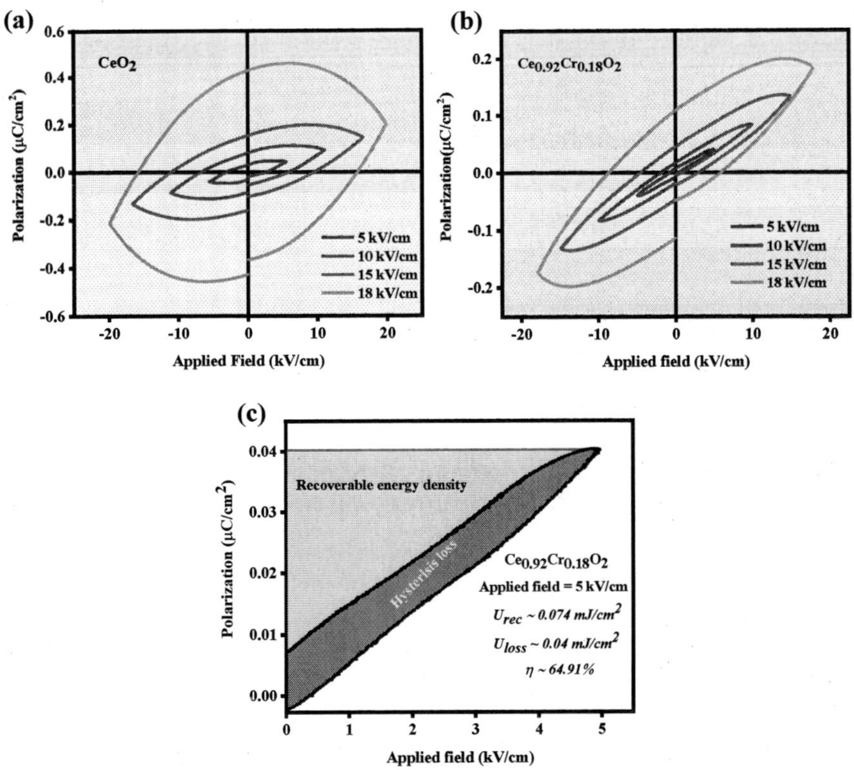

Figure 5: (a-b) P-E Loop of the Pristine CeO_2 and Cr-doped CeO_2 Compounds, (c) Graphical Representation of Recoverable Energy, Hysteresis Loss of Cr Doped CeO_2 Compound

calculation of energy density and energy efficiency of the samples using a monopole is illustrated in Figure 5(c). The light green and red segment of the monopole graph represent the recoverable energy density and hysteresis loss for the $Ce_{1-x}Cr_xO_2$ (x = 18%) sample. The recoverable energy density is found to be 0.074 mJ/cm², while the corresponding hysteresis loss is 0.04 mJ/cm², yielding an overall energy-storage efficiency of 64.91%. The efficiency is calculated using the formula $\eta = U_{rec}/(U_{rec} + U_{loss})$, indicating that nearly two-thirds of the input electrical energy can be effectively recovered. In contrast, the remaining fraction is dissipated due to irreversible domain switching and conduction-related losses. Although the recoverable energy density is modest, the efficiency value highlights the potential of this system for energy-storage applications, with further improvements expected through microstructural optimization, defect control, and enhancement of breakdown strength. The calculated parameters are listed in Table 3.

Table 3: Calculated Recoverable Energy Density, Hysteresis Loss and Efficiency of the $Ce_{1-x}Cr_xO_2$ Compounds

Samples	*Applied Field (kV/cm)*	*Recoverable Energy Density (U_{rec}) mJ/cm²*	*Hysteresis Loss (U_{loss}) mJ/cm²*	*Efficiency (η%)*
CeO_2	5	0.016	0.210	7.11
$Ce_{0.82}Cr_{0.18}O_2$	5	0.074	0.040	64.91

V. Conclusion

The Cr-doped CeO_2 (X = 0% and 18%) compounds were synthesied via the solid-state route. The XRD analysis confirmed the cubic fluorite structure with a defined crystallite size of 65 nm, retaining the microstrain of 1.09×10^{-3}. The FESEM images reveal the porousity in between the grains, along with the average grain size range of 0.20–0.25 μm. The dielectric constant of Cr-doped CeO_2 is high due to the formation of a large space charge within the sample, indicating a promising candidate for energy storage applications. The negative loss at room temperature at a lower frequency represents the low leakage current. The leakage current reduces more at the Cr-doped CeO_2 compound as a result the ac conductivity is less compared to the CeO_2 material. The ferroelectric response show improvement at Cr doped CeO_2 compound, with the high recoverable energy density of 0.074 mJ/cm² at 5 kV/cm applied field, with the efficiency of 64.91% which highlights the scope for the energy storage applications.

Acknowledgement

R. Brahma gratefully acknowledges the financial support for the project under the 'Empowerment and Equity Opportunities for Excellence in Science' from DST-SERB India, (sanction order number EEQ/2023/000247). R. Basumatary acknowledges the fellowship support received through this project. RB also expresses sincere gratitude for the assistance received from INUP-i2i, the Central Instrument Facility at IIT Guwahati, and IASST Guwahati for providing essential analytical resources for this study. The authors further thank the Department of Physics, IIT Guwahati for the ferroelectric property measurements.

References

[1] R. Basumatary, S. Sahu, D. Goyari, D. Pamu and R. Brahma, Effect of manganese (Mn) doping on the structural, dielectric, ferroelectric, and ferromagnetic properties of ceria (CeO_2), *J. Alloys Compd.*, 999, 174958, (2024).

[2] K.O. Abdulwahab, M.M. Khan and J.R. Jennings, Doped Ceria Nanomaterials: Preparation, Properties, and Uses, ACS Omega, 8/34, 30802–30823, (2023).

[3] Y.P. Xiong, H. Kishimoto, K. Yamaji, M. Yoshinaga, T. Horita, M.E. Brito and H. Yokokawa, Electronic conductivity of pure ceria, *Solid State Ionics,* 192/1, 476–479 (2011).

[4] R.V. Mangalaraja, B.K. Chandrasekhar and P. Manohar, Effect of ceria on the physical, mechanical and thermal properties of yttria stabilized zirconia toughened alumina, *Mater. Sci. Eng. A,* 343/1–2, 71–75 (2003).

[5] R. Maurya, A. Gupta, S. Omar and K. Balani, Effect of sintering on mechanical properties of ceria reinforced yttria stabilized zirconia, *Ceram. Int.,* 42/9, 11393–11403 (2016).

[6] N. Momin, J. Manjanna, S. Kobayashi, S.T. Aruna, S.S. Kumar and G.P. Nayaka, Synthesis and ionic conductivity of calcium-doped ceria relevant to solid oxide fuel cell applications, *Mater. Adv.,* 33/12, 8780–8791 (2022).

[7] Y. Zhang, J. Liu, M. Singh, E. Hu, Z. Jiang, R. Raza, F. Wang, J. Wang, F. Yang and B. Zhu, Superionic Conductivity in Ceria-Based Heterostructure Composites for Low-Temperature Solid Oxide Fuel Cells, *Nano-Micro Lett.,* 12/1, 1–20 (2020).

[8] S. Omar, Doped Ceria for Solid Oxide Fuel Cells, *Cerium Oxide - Appl. Attrib.,* 2019.

[9] B. Krishnamurthy and H. Ramasubramanian, Modelling the leakage current in a solid oxide fuel cell with bi-layer electrolyte, *J. Electrochem. Sci. Eng.,* 10/3, 293–304 (2020).

[10] S. Ghosh and B. Rani, Cr-doped ceria for solid-state symmetric supercapacitor using gel-electrolyte, *Ionics (Kiel).,* 31/6, 6545–6552 (2025).

[11] B.M. Harish, B.S. Avinash, V.S. Chaturmukha, H.S. Jayanna, A.R. Lamani and S. Yallappa, Synthesis and dielectric properties of Cr-substituted CeO_2 nanoparticles, *J. Mater. Sci. Mater. Electron.*, 29/9, 7402–7411 (2018).

[12] F. Pedroli, A. Marrani, M.Q. Le, O. Sanseau, P.J. Cottinet and J.F. Capsal, Reducing leakage current and dielectric losses of electroactive polymers through electro-annealing for high-voltage actuation, *RSC Adv.*, 9/23, 12823–12835 (2019).

[13] E. Miura-Stempel, A.G. Oregon, S.M. Harvey, J.J. De Yoreo, C.L. Chen and B.M. Cossairt, CeO_2 Nanoparticle Doping as a Probe of Active Site Speciation in the Catalytic Hydrolysis of Organophosphates, *ACS Appl. Nano Mater.*, 7/13, 15498–15507 (2024).

[14] S.D. Neelapala and H. Dasari, Catalytic soot oxidation activity of Cr-doped ceria ($Ce_{1-x}Cr_xO_{2-\delta}$) synthesized by sol-gel method with organic additives, *Mater. Sci. Energy Technol.*, 1/2, 155–159 (2018).

[15] Basumatary, R., Kalita, P., Bailung, H. and Brahma, R., Facile synthesis of CeO_2 nanoparticles through plasma-liquid interaction, *Materialia*, 39, 102367 (2025).

[16] R. Basumatary, S. Sahu, D. Pamu and R. Brahma, Exploring the impact of Cr ions on CeO_2-based material: Dielectric, ferroelectric, and ferromagnetic properties for multifunctional device applications, *J. Alloys Compd.*, 1041, (2025).

[17] S. Sharma, G. Rojas-Geoge, M. Kumar, P. Kumar, R. Kumar, S. Kumar, J.M. Siqueiros, and O.R. Herrera, Impact of Cr Doping on the Structural, Optical, and Magnetic Properties of Sol-Gel-Synthesized $Bi_{0.80}Ba_{0.10}Pr_{0.10}FeO_3$ Nanopowders, *ACS Omega*, 9/25, 27549–27558 (2024).

[18] S.Y. Chen, K.W. Fong, T.T. Peng, C.L. Dong, A. Gloter, D.C. Yan, C.L. Chen, H.J. Lin and C.T. Chen, Enhancement of ferromagnetism in CeO_2 nanoparticles by nonmagnetic Cr^{3+} doping, *J. Phys. Chem. C*, 116/50, 26570–26576 (2012).

[19] E. Axelrod, A. Puzenko, Y. Haruvy, R. Reisfeld and Y. Feldman, Negative dielectric loss phenomenon in porous sol-gel glasses, *J. Non. Cryst. Solids*, 352/40-41 4166–4173 (2006).

[20] J.A. Bartkowska and D. Bochenek, Microstructure and dielectric properties of BF–PFN ceramics with negative dielectric loss, *J. Mater. Sci. Mater. Electron.*, 29/20, 17262–17268 (2018).

[21] A. Azam, A.S. Ahmed, M. Chaman and A.H. Naqvi, Investigation of electrical properties of Mn doped tin oxide nanoparticles using impedance spectroscopy, *J. Appl. Phys.*, 108/9, (2010).

[22] B. Veena and M. Seetha, Synthesis and Characterization of Cr doped CeO_2 nanoparticles for Rhodamine B dye degradation, *Heliyon*, 10/16, (2024).

CHAPTER 9

Nanomaterials in the Environment: Applications, Risks, and Future Directions

Aziz Hussain[1,2*], Ranjit Kakati[1] and Raju Ojah[1]

[1]DBT-NER Advanced Level Institutional Biotech Hub, Chaiduar College (Autonomous), Gohpur, Assam, India

[2]Department of Zoology, Cotton University, Panbazar, Guwahati, Assam, India

*E-mail: biotechcdc23@gmail.com

ABSTRACT: Nanomaterials (NMs) have transformed the environmental sciences by providing inventive opportunities to create technologies to mitigate pollution, monitoring resources, and creating sustainable solutions. The use of NMs offers tangible benefits for environmental protection through water purification, soil detoxification, and pollutant detection. It is obvious that the promise of nanotechnology has opportunities in environmental preservation as well as potential risks (i.e. release into their environment (e.g. bioaccumulation and toxicological effects seen with both aquatic and terrestrial organisms; inadequate tools for risk assessment long-term impacts). This chapter focuses on promising developments in the remediation use of NMs, mechanisms by which they may produce negative consequences, and the need to establish regulatory lower thresholds to ensure both the safety and monitoring of NMs. Finally, considerations of future research need such as safer design for NMs, implementation for broader use to increase benefits and limit risks are discussed.

Keywords: Nanomaterials (NMs), Environment, Pollution, Nanotechnology.

I. INTRODUCTION

Nanomaterials (NMs) are defined as materials with one or more dimensions in the range of 1–100 nm, making them one of the most important technological advances in the 21st century [1]. At the nano-scale, NMs possess different physical, chemical and biological properties compared to bulk materials, meaning they have important new characteristics such as increased reactivity, unique optical properties, and improved mechanical properties [2]. This combination of properties has accelerated the development of nanotechnology and facilitated the use of NMs in consumer applications, industrial processing, and medicine.

Although these new characteristics offer great value, NMs also raise risks to potential environmental consequences once they are released into the environment. NMs, similar to other pollutants, may be unintentionally introduced into ecosystems from being produced, used, or disposed of [3]. This article assesses both the advantages of NMs and the potential environmental risks associated with them. It explores significant applications, report and evaluate environmental concerns, and outline directions for the safe and sustainable development of NMs.

II. Classification and Properties of Nanomaterials

NMs are a highly diverse group of materials. They can be classified based on several criteria, most commonly their dimensionality, composition, and origin.

A. Classification by Dimensionality

This scheme categorizes NMs based on the number of dimensions that are not confined to the nanoscale.

- **Zero-Dimensional (0D) Nanomaterials:** All three dimensions are within the nanoscale (typically <100 nm). They are often referred to as quantum dots or nanospheres. Examples: Spherical nanoparticles of gold (AuNPs), silver (AgNPs), and semiconductor quantum dots (e.g., Cadmium Selenide, CdSe) [4].
- **One-Dimensional (1D) Nanomaterials:** Two dimensions are in the nanoscale, while one is extended. This results in elongated structures. Examples: Carbon nanotubes (CNTs), nanowires, nanofibers, and nanorods [5].
- **Two-Dimensional (2D) Nanomaterials:** Only one dimension is in the nanoscale, resulting in sheet-like structures. Examples: Graphene, nanosheets [e.g., molybdenum disulfide (MoS_2)], and nanoclays [6].
- **Three-Dimensional (3D) Nanomaterials:** These are bulk materials that possess a nanostructured architecture or features, such as nanocomposites or materials with nanoscale porosity. They are not typically considered nanoparticles but are part of the broader family of NMs. Examples: Dendrimers, metal-organic frameworks (MOFs), and nanocomposites [7].

B. Classification by Composition

This is the most common classification method, categorizing NMs by the material they are made from.

Table 1: Classification of Nanomaterials by Composition

Category	*Sub-Category*	*Description and Examples*
Carbon-Based	Fullerenes, CNTs, Graphene	Composed primarily of carbon, these materials exhibit exceptional strength and electrical conductivity. (e.g., C_{60}, SWCNTs, MWCNTs) [5].
Inorganic-Based	Metal NMs	Pure metal nanoparticles known for their optical and electronic properties. (e.g., AuNPs, AgNPs, Platinum NPs) [4].
	Metal Oxide NMs	Offer unique catalytic, magnetic, and semiconducting properties. (e.g., TiO_2, ZnO, Fe_3O_4) [8].
	Semiconductor NMs	Exhibit size-tunable optical and electronic properties (quantum confinement). (e.g., CdSe, CdS, ZnS-often called quantum dots) [9].
Organic/Polymer-Based	Dendrimers, Micelles, Nanoshells	Made from organic polymers. Widely used in biomedicine for drug delivery due to their biocompatibility [10].
Lipid-Based	Liposomes, Solid Lipid NMs (SLNs)	Spherical vesicles composed of a lipid bilayer, primarily used for encapsulating and delivering drugs and vaccines [11].
Composite-Based	Nanocomposites	Multiphase materials where nanoparticles are embedded within a matrix (e.g., polymer, metal, or ceramic) to enhance its properties.

III. Properties of Nanomaterials

A. Physical Properties

The exceptional physical properties of NMs arise essentially from their very small size (1-100 nanometres) and their large ratio of surface area to volume [12]. This increases the surface area, increases the surface energy and the tendency for a particle to agglomerate, and the melting point is lower than the bulk material [13]. The morphology and size of these materials can also be controlled precisely, which leads to properties that have distinct morphological and structural aspects [12].

B. Chemical Properties

NMs are significantly more chemically reactive than a bulk material. The ratio of atoms that are exposed on the surface of the particle accounts for a very significant fraction of atoms in NMs [12]. This causes an increase in catalytic performance and surface activity. It is also possible to tune the chemical composition and surface chemistry of functionalizing the surface of the nanoparticle to include zeta potential and photocatalytic properties [4].

C. Electrical Properties

At the nanoscale, the electrical properties would be affected by quantum confinement effects due to changes in the band structure that lead to discrete electronic states [4]. Although materials like carbon nanotubes and fullerenes have excellent conductivity and even act as conductors or semiconductors [12].

D. Optical Properties

Due to the small size and special structure of NMs, a variety of their optical properties can be observed. To illustrate, gold nanoparticles present different colors depending on their size and shape. Moreover, the capacity of semiconductor NMs to absorb light and fluoresce gets lower as their size decreases [4]. Such properties make them approachable in imaging, sensing, and photonic devices [12].

E. Magnetic Characteristics

The magnetic properties of the same materials at the nanoscale can be very different. One example of such materials that can exhibit superparamagnetism is magnetic nanoparticles. Even a few atoms of gold, which is a non-magnetic material, under certain conditions can show magnetic properties [13]. The magnetic NMs are the components used in MRI, the vehicle for targeted delivery, and magnetic fluid hyperthermia [12].

F. Mechanical Properties

NMs typically retain or even enhance their structural properties of strength, hardness, and elasticity through the refinement of their grain boundaries and quantum effects at the nanoscale. For instance, carbon nanotubes are more rigid and lighter than regular materials by a large margin [12].

G. Thermal Properties

They are mostly featured with high thermal conductivity and stability compared to large materials, which enable easier heat transfer and dissipation. These points are great, for example, in the production of microelectronics and composite materials [12].

IV. Environmental Applications of Nanomaterials

A. Water Treatment

NMs' applications in the treatment of water are perhaps the most visible of their many applications, with nanoparticles such as zero-valent iron, titanium dioxide, and graphene oxide having been used for the removal of heavy metals, organic pollutants, and pathogens from water systems using physical processes such as adsorption, photocatalysis, and membrane filtration [14, 15]. Nanomaterial membranes enhance filtration by decreasing pore size and increasing rejection of contaminants [16].

B. Soil Remediation

Soil decontamination via immobilization and degradation of organic contaminants and heavy metals is a promising application of nanotechnology. To this end, nano-iron and nanoclays have been shown to be effective for contaminant removal, while enhancing plant pollutant removal is provided through nanophytoremediation techniques [17].

C. Air Pollution Control

The use of photocatalysts such as TiO_2 and specific coatings of NMs has become increasingly popular for air pollution control and self-cleaning surfaces, as many NMs break down volatile organic compounds and nitrogen oxides [18].

D. Environmental Monitoring

Biosensors and nanosensors that employ quantum dots, carbon nanotubes, or metal nanoparticles demonstrate improved sensitivity and selectivity for environmental contaminants at trace levels [19].

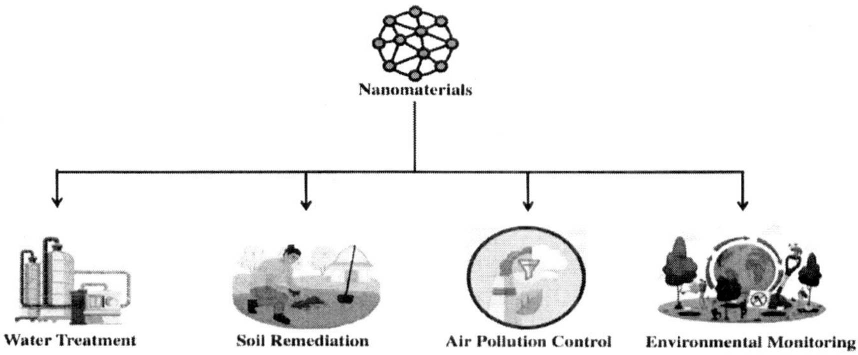

Figure 1: Different Applications of Nanomaterials in Environment

V. ENVIRONMENTAL RISKS AND NANOTECHNOLOGY

The unique attributes of engineered nanomaterials (ENMs) pose unique risks. Nanotoxicology is the study of the negative consequences of ENMs on living organisms and systems. The main worry with ENMs is that their small size allows them to cross biological barriers (i.e., the cell membranes and the blood-brain barrier) that larger particles cannot pass, which can lead to previously unknown toxic effects [20].

A. Fate and Behavior in the Environment

NMs find their way into nature through various routes, for example, the release of the industry, the wearing down of consumer products, and the spillage of the area [21]. In their tiny sizes, large surface areas, and high reactivities, they carry the behavior of complex systems such as aggregation, dissolution, transformation, and change of bioavailability, which are determined by parameters such as pH, ionic strength, and organic matter content [22, 23]. The changes they make to the environment influence their mobility and life span in it, which may result in their accumulation in living organisms and the passing on of food chains within ecosystems [24].

B. Ecotoxicological Effects

NMs have the ability to cause harmful effects to various living beings and to the food chains that they are part of. The aquatic organisms, to name but a few, algae, daphnia, and fish, have shown in numerous experiments that they are vulnerable to the effects of silver, zinc oxide, and titanium dioxide nanoparticles, resulting in such conditions as oxidative stress, membrane damage, and reproductive failure [25, 26]. The soil dwellers, such as earthworms and microbes, are also not spared from the toxic effects, which, in turn, cause nutrient cycling in the soil to be disrupted and the health of the organisms that live in the soil to deteriorate [27]. The materials terrestrial plants may become after exposure to NMs include the changes in seed germination, plant retardation, and genotoxicity [28].

C. Human and Mammalian Health Considerations

The possible ways for humans to get in touch with these substances are through breathing, skin touching, and eating, as these particles will possibly reach cellular levels and are later spread over the body [29]. Several pieces of research show that there can be inflammatory reactions in the lungs, cell death, and DNA damage, which are caused by the chronic exposure to certain NMs; however, to date, only the short-term health impacts of such exposures have been verified, and the long-term health effects are still to be discovered [30].

D. Regulatory and Risk Assessment Challenges

The present regulatory systems for nanotechnologies have shortcomings in coping with NMs' rapid technological development without clear directions for how to carry out standardized tests specifically for NMs. The risks involved with NMs are hard to figure out because the properties of NMs in terms of physics, chemistry, and effects on the environment are hardly

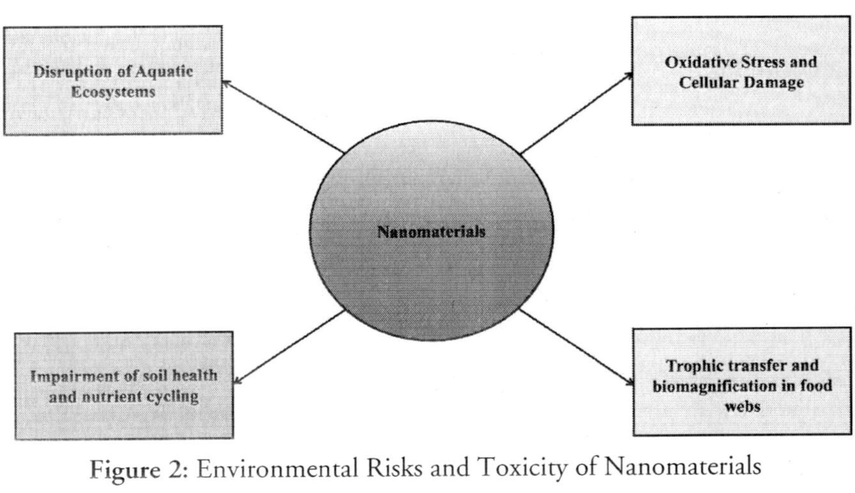

Figure 2: Environmental Risks and Toxicity of Nanomaterials

constant and also present in various combinations [21]. It is necessary to adopt a holistic approach that combines exposure assessment, hazard identification, and predictive toxicology in order to solve this problem [31].

VI. Future Directions Towards Sustainable Nanotechnology

Nanotechnology is a key goal today to the point that researchers are making this a science to create safer and more eco-friendly materials and processes. Progress is gradually shifting towards the creation of less toxic NMs that disintegrate faster in the natural environment [32]. The use of green synthesis methods that employ biological agents and renewable resources mitigates the release of harmful byproducts and also reduces the energy required for the production of NMs [33, 34].

More thorough life cycle assessments and environmental monitoring have made it possible to trace the fate of NMs and their impacts, which in turn facilitates the setting of regulations and risk management [19, 35]. Responsible progress depends on different areas working together, having well-defined rules, and the public's involvement. This method enables nanotechnology to deliver enduring advantages and be in line with a circular economy that envisages NMs as being recyclable, disintegrating rapidly, and having a negligible environmental impact. These are the provisions for the creation of the future safe and efficient nanotechnologies.

VII. Conclusion

NMs can be a great source of environmental innovation and can be used as technology for clean and renewable energy, air purification, and water purification. With their nanoscale properties, they can be highly efficient and reactive, which makes them very useful in resolving global challenges caused by pollution and resource scarcity. On the other hand, their most outstanding features can also cause them to be quite a puzzle in terms of potential environmental and health risks as their interactions in ecosystems and biological systems are not the latest scientific studies that confirm it. Thus, securing safe and sustainable advancement in nanotechnology calls for incorporating safety into every stage of research and development. Their use should be accompanied by detailed risk assessments, constant monitoring, and flexible regulatory frameworks. Collaboration among scientists, industry stakeholders, policymakers, and the public to have a common sense of responsibility, openness, and confidence as this revolutionary area is going on is as important as the above-mentioned measures.

References

[1] C. Buzea, I.I. Pacheco and K. Robbie (2007). NMs and nanoparticles: Sources and toxicity. Biointerphases, 2(4), MR17–MR71.

[2] A. Nel, T. Xia, L. M¢Dler and N. Li (2006). Toxic potential of materials at the nanolevel. Science, 311(5761), 622–627.

[3] S.J. Klaine, P.J.J. Alvarez, G.E. Batley, T.F. Fernandes, R.D. Handy, D.Y. Lyon, S. Mahendra, M.J. McLaughlin and J.R. Lead (2008). NMs in the environment: Behavior, fate, bioavailability, and effects. Environmental Toxicology and Chemistry, 27(9), 1825–1851.

[4] M.-C. Daniel and D. Astruc (2004). Gold nanoparticles: Assembly, supramolecular chemistry, quantum-size-related properties, and applications toward biology, catalysis, and nanotechnology. Chemical Reviews, 104(1), 293–346.

[5] S. Iijima (1991). Helical microtubules of graphitic carbon. Nature, 354(6348), 56–58.

[6] K.S. Novoselov, A.K. Geim, S.V. Morozov, D. Jiang, Y. Zhang, S.V. Dubonos, I.V. Grigorieva and A.A. Firsov (2004). Electric field effect in atomically thin carbon films. Science, 306(5696), 666–669.

[7] D.A. Tomalia, H. Baker, J. Dewald, M. Hall, G. Kallos, S. Martin, J. Roeck, J. Ryder and P. Smith (1985). A new class of polymers: Starburst-Dendritic macromolecules. Polymer Journal, 17(1), 117–132.

[8] A.K. Gupta and M. Gupta (2004). Synthesis and surface engineering of iron oxide nanoparticles for biomedical applications. Biomaterials, 26(18), 3995–4021.

[9] A.P. Alivisatos (1996). Semiconductor clusters, nanocrystals, and quantum dots. Science, 271(5251), 933–937.

[10] S. Svenson and D. Tomalia (2005). Dendrimers in biomedical applications—reflections on the field. Advanced Drug Delivery Reviews, 57(15), 2106–2129.

[11] V.P. Torchilin (2005). Recent advances with liposomes as pharmaceutical carriers. Nature Reviews Drug Discovery, 4(2), 145–160.

[12] N. Baig, I. Kammakakam and W. Falath (2021). NMs: a review of synthesis methods, properties, recent progress, and challenges. Materials Advances, 2(6), 1821–1871.

[13] B. Mekuye and B. Abera (2023). NMs: An overview of synthesis, classification, characterization, and applications. Nano Select, 4(8), 486–501.

[14] A. Roy, A. Sharma, S. Yadav, L.T. Jule and R. Krishnaraj (2021). NMs for remediation of environmental pollutants. Bioinorganic Chemistry and Applications, 2021, 1–16.

[15] N. Chausali, J. Saxena and R. Prasad (2023). Nanotechnology as a sustainable approach for combating the environmental effects of climate change. Journal of Agriculture and Food Research, 12, 100541.

[16] M.L. Del Prado-Audelo, I.G. Kerdan, L. Escutia-Guadarrama, J.M. Reyna-González, J.J. Magaña and G. Leyva-Gómez (2021). Nanoremediation: NMs and nanotechnologies for environmental cleanup. Frontiers in Environmental Science, 9.

[17] A.R. Dhanapal, M. Thiruvengadam, J. Vairavanathan, B. Venkidasamy, M. Easwaran and M. Ghorbanpour (2024). Nanotechnology approaches for the remediation of agricultural polluted soils. ACS Omega, 9(12), 13522–13533.

[18] M.I. Sohail, A.A. Waris, M.A. Ayub, M. Usman, M.Z.U. Rehman, M. Sabir and T. Faiz (2019). Environmental application of NMs: A promise to sustainable future. In Comprehensive analytical chemistry (pp. 1–54).

[19] F. Zhang, Z. Wang, W.J.G.M. Peijnenburg and M.G. Vijver (2022). Review and prospects on the ecotoxicity of mixtures of nanoparticles and hybrid NMs. Environmental Science & Technology, 56(22), 15238–15250.

[20] H.S. Sharma, S.F. Ali, S.M. Hussain, J.J. Schlager and A. Sharma (2009). Influence of Engineered Nanoparticles from Metals on the Blood-Brain Barrier Permeability, Cerebral Blood Flow, Brain Edema and Neurotoxicity. An Experimental Study in the Rat and Mice Using Biochemical and Morphological Approaches. Journal of Nanoscience and Nanotechnology, 9(8), 5055–5072.

[21] F. Gottschalk, T. Sonderer, R.W. Scholz and B. Nowack (2009). Modeled environmental concentrations of engineered NMs (TIO2, ZNO, AG, CNT, fullerenes) for different regions. Environmental Science & Technology, 43(24), 9216–9222.

[22] L.K. Limbach, P. Wick, P. Manser, R.N. Grass, A. Bruinink and W.J. Stark (2007). Exposure of engineered nanoparticles to human lung epithelial cells: influence of chemical composition and catalytic activity on oxidative stress. Environmental Science & Technology, 41(11), 4158–4163.

[23] E. Navarro, A. Baun, R. Behra, N.B. Hartmann, J. Filser, A. Miao, A. Quigg, P.H. Santschi and L. Sigg (2008). Environmental behavior and ecotoxicity of engineered nanoparticles to algae, plants, and fungi. Ecotoxicology, 17(5), 372–386.

[24] C.V. Kumar, V. Karthick, V.G. Kumar, D. Inbakandan, E.R. Rene, K.U. Suganya, A. Embrandiri, T.S. Dhas, M. Ravi and P. Sowmiya (2022). The impact of engineered NMs on the environment: Release mechanism, toxicity, transformation, and remediation. Environmental Research, 212(Pt B), 113202.

[25] J. Fabrega, S.N. Luoma, C.R. Tyler, T.S. Galloway and J.R. Lead (2010). Silver nanoparticles: Behaviour and effects in the aquatic environment. Environment International, 37(2), 517–531.

[26] X. Zhao, J. Liang, Y. Li, C.W. Isaacson, X. Meng, Y. Luo, X. Zhou, H. Hollert, Q. Zhou, R. Liu, Z. Cai and G. Jiang (2025). Trophic transfer of metal nanoparticles in an aquatic food chain diminishes their toxicity disparities. Environmental Science & Technology, 59(13), 6812–6824.

[27] K. Tiede, A.B. Boxall, S.P. Tear, J. Lewis, H. David and M. Hassellöv (2008). Detection and characterization of engineered nanoparticles in food and the environment. Food Additives & Contaminants Part A, 25(7), 795–821.

[28] M.V. Khodakovskaya, K. De Silva, A.S. Biris, E. Dervishi and H. Villagarcia (2012). Carbon nanotubes induce growth enhancement of tobacco cells. ACS Nano, 6(3), 2128–2135.

[29] G. Oberdörster, E. Oberdörster and J. Oberdörster (2005). Nanotoxicology: An Emerging Discipline Evolving from Studies of Ultrafine Particles. Environmental Health Perspectives, 113(7), 823–839.

[30] C.M. Sayes, K.L. Reed and D.B. Warheit (2011). Nanoparticle toxicology: Measurements of pulmonary hazard effects following exposures to nanoparticles. Methods in Molecular Biology, 726, 313–324.

[31] S. Gottardo, A. Mech, J. Drbohlavová, A. Małyska, S. Bøwadt, J.R. Sintes and H. Rauscher (2021). Towards safe and sustainable innovation in nanotechnology: State-of-play for smart NMs. NanoImpact, 21, 100297.

[32] S.K. Kodadi, D. Manimekalai, D. Rabeek, D.U. Kalme, P, S. and D.K. Nathsharma (2025). Nanotechnology and Green NMs: modern eco- friendly sustainable approach for combating the environmental effects of climate change. International Journal of Environmental Sciences, 11(12s), 238–254.

[33] D. Gupta, A. Boora, A. Thakur and T.K. Gupta (2023). Green and sustainable synthesis of NMs: Recent advancements and limitations. Environmental Research, 231(Pt 3), 116316.

[34] R. Sharma and A. Tripathi (2021). Green synthesis of nanoparticles and its key applications in various sectors. Materials Today Proceedings, 48, 1626–1632.

[35] C. Gambardella and A. Pinsino (2022). Nanomaterial ecotoxicology in the Terrestrial and Aquatic environment: A Systematic review. Toxics, 10(7), 393.

Chapter 10

Nanomaterials and Material Innovation for Air Quality Management and Climate Sustainability

Parminder Kaur*, Iqbal Singh and Manoj Kumar
Department of Physics, Akal University, Talwandi Sabo, India
*E-mail: parminder_phy@auts.ac.in

ABSTRACT: Materials innovation for mitigating air pollution and climate change is a critical global concern that requires multidisciplinary solutions which integrate atmospheric science with advanced materials. The urgency is crucial as 2024 marked the first year to exceed 1.5ºC above pre-industrial levels and atmospheric carbon dioxide (CO_2) concentrations reaching 422.5 ppm (50% higher than pre-industrial levels). This chapter investigates the role of advanced materials in monitoring, mitigating, and managing atmospheric pollutants, including CO_2, particulate matter 2.5 ($PM_{2.5}$), and greenhouse gases. Advanced materials provide measurable solutions such as metal organic frameworks (MOFs) exhibit CO_2 capture capacities surpassing 15–20 wt%, whereas photocatalytic nanocomposites attain pollutant degradation efficiencies of 80–95%. Smart sensor networks provide real time monitoring with parts per billions level detection limits. This chapter particularly focuses on 1D and 2D nanomaterials as they are used in efficient photovoltaic devices, integrating with advances in perovskite silicon tandem solar cells that now surpass the 30% efficiency limit. When strategically implemented in the urban regions, these technologies deliver measurable outcomes mainly 20–40% reduction in $PM_{2.5}$ concentrations and 15–25% air quality improvement. The analysis of cost-effectiveness indicates that material-based systems can achieve a reduction in pollutants at a rate of 50–150 per ton of CO_2 equivalent, while manufacturing costs are expected to decline by 30–50% due to scalable production. The life cycle assessments also indicate 60–80% decline in environmental impact as compared to the conventional technologies. By connecting innovation in material science with atmospheric sustainability, we can improve public health outcomes, reduce environmental risks, and build a climate resilient society transition toward Intergovernmental Panel on Climate Change (IPCC) targets.

Keywords: Nanomaterials, Solar Energy, Pollution Control, Sustainable Materials Innovation, Climate Resilience.

I. Introduction

Air pollution and climate change represent the pressing global crises of the 21st century, with profound environmental, health, social, and economic impacts. According to the World Health Organization (WHO) [1], over 99% of the global population breathes polluted air, where fine particulate matter (PM) and ground-level ozone (O_3) are the leading contributors [2]. The Lancet Commission estimates that air pollution is responsible for approximately 7 million premature deaths worldwide each year, positioning it as the foremost environmental risk factor for mortality [3, 4]. Simultaneously, the Intergovernmental Panel on Climate Change (IPCC) [5] reports that global average temperature already has crossed the 1.5 °C threshold above pre-industrial levels for the first time in 2024, thereby increasing the urgency for mitigation efforts. [5, 6].

The interconnections of atmospheric pollutants highlight the complexity of challenges as represented in Figure 1. Pollutants exhibit divergent climate forcing behaviors; some are contributing to warming (Carbon dioxide (CO_2), Methane (CH_4), Tropospheric O_3, Black Carbon (BC)), others to cooling (Sulfur dioxide (SO_2)), and some displaying mixed effects (Nitrogen Oxide (NO_x), Particulate Matter ≤2.5 µm ($PM_{2.5}$)), while also imposing severe health burdens [7, 8]. Greenhouse gases (GHGs) like CO_2, CH_4, and Nitrous Oxide (N_2O) drive long term warming, with atmospheric CO_2 exceeding 420 ppm in 2024 [9]. Short lived climate pollutants (SLCPs), such as BC and tropospheric O_3, exacerbate global warming and worsen air quality [10, 11]. $PM_{2.5}$ serves as both an air pollutant and a climate forcing agent through direct radiation effects and cloud property modifications as it acts as cloud condensation nuclei (CCN) [12]. The interrelated nature of these pollutants reflects policies targeting one pollutant can deliver co-benefits across climate and health domains [13, 14].

Conventional mitigation approaches, such as fossil fuel (FF) reduction, pollutant emission controls, and clean energy transitions, are necessary yet insufficient when considered in isolation [15, 16]. Therefore, they require complementary technological innovations operating at the material scale, offering efficient pathways. Advanced materials facilitate the degradation of toxic gases by photocatalysts under sunlight [17], porous frameworks capturing GHGs with high selectivity [18], and nanomaterials form the backbone of next-generation sensors and energy devices [19, 20]. Photocatalytic pavements enhance air quality and reduce heat island effects [21], whereas nanomaterial-based sensors offer real time data for responsive policy measures [22].

The present chapter explores how materials innovation contributes to monitoring, mitigating, and managing air pollution in a climate constrained world. By linking atmospheric science with material design, it demonstrates how targeted innovations can reduce environmental risks, protect public health, and advance global sustainability targets.

II. ATMOSPHERIC POLLUTANTS AND CLIMATE LINKAGES

The complex relationship between atmospheric pollution and climate change is one of the pressing environmental problems of present day. Atmospheric pollutants harm air quality, pose serious health hazards, and fundamentally affect energy balance of Earth via radiative forcing processes. Therefore, it becomes critical to understand these relationships for establishing successful mitigation methods that meet both air quality and climate goals.

A. Particulate Matter ($PM_{2.5}$ and PM_{10})

PM, categorized as $PM_{2.5}$ (particles ≤2.5 μm) and PM_{10} (particles ≤10 μm), originates from combustion of FFs, biomass burning (BB), and industrial processes, which leads to the secondary formation via atmospheric reactions of SO_2, NO_x, and volatile organic compounds (VOCs) [23]. $PM_{2.5}$ particles penetrate deep into the lungs and circulatory systems, causing cardiovascular disease and premature mortality. $PM_{2.5}$ contributes to 7 million annual deaths worldwide as estimated by WHO [1]. PM affects Earth's radiative balance through direct scattering and absorption of solar radiation, where net effects depend upon the composition [24].

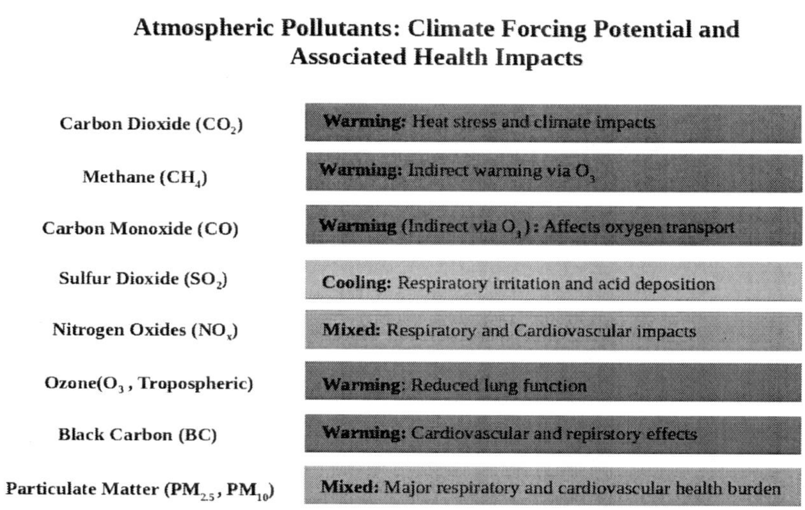

Figure 1: Climate Forcing Potential and Health Impacts of Major Atmospheric Pollutants

B. Greenhouse Gases (GHGs)

GHGs regulate Earth's climate by trapping outgoing infrared radiation and amplifying the greenhouse effect. CO_2, which exceeded 420 ppm in 2024, remains the primary GHG due to its long atmospheric lifetime i.e., 300–1000 years, and significant warming effect [25]. CH_4, with a 100-year global warming potential (GWP), 28–34 times that of CO_2, emitted mainly from agriculture, FF extraction, and landfills. Fertilizer-induced NO_x emission has a GWP of 265–298 and a 121-year lifetime. BC is emitted from incomplete combustion, links air pollution and climate forcing by inducing strong radiative effects.

C. Short Lived Climate Pollutants (SLCPs)

SLCPs are a class of atmospheric pollutants with relatively short atmospheric lifetimes having significant impacts on climate change and air quality. Ground level O_3, formed through photochemical reactions, involving NO_x and VOCs, acts as both a harmful air pollutant and a GHG. Other important SLCPs include BC, which is often co-emitted with cooling aerosols such as sulfates, resulting in mixed radiative forcing effects (both direct and indirect). Regional climate warming of 0.75–1.25°C solely due to BC emissions has been documented [26].

D. Pathways of Pollutant Interaction with Climate Systems

Atmospheric pollutants directly affect Earth's energy balance by modifying the absorption, scattering, and emission of radiation [27]. In addition to direct radiative impacts, pollutants influence climate via feedback mechanisms. The aerosol-cloud interactions modify cloud micro physics, which affects the precipitation patterns. The deposition of absorbing aerosols on ice and snow decreases albedo, thereby accelerating melting. Pollutants influence bio-geochemical cycles e.g., nitrogen deposition may promote plant growth but also leads to eutrophication, ozone exposure harms vegetation, diminishing photosynthetic capacity and carbon uptake.

Figure 2 shows that long-lived greenhouse gases are the main cause of positive radiative forcing. CO_2 contributes 1.66 W/m^2, followed by CH_4, and N_2O, while tropospheric ozone adds 0.35 W/m^2 [28]. On the other hand, aerosols exhibit cooling effects via direct scattering and cloud albedo modifications, BC on snow contributes a warming effect of 0.1 W/m^2. The net anthropogenic radiative forcing currently exceeds 1.6 W/m^2, with the average radiative forcing of 22 long-lived greenhouse gases continuously monitored by

National Oceanic and Atmospheric Administration (NOAA), which provides essential information for climate model validation and policy development [25].

Table 1: Key Atmospheric Pollutants, Sources, Health Impacts, and Climate Forcing Potential

Pollutant	*Primary Sources*	*Health Impacts*	*Atmospheric Lifetime*	*Radiative Forcing (W/m^2)*	*Climate Effect*
CO_2	FF combustion, deforestation, cement production	Indirect (climate-related)	300–1000 years	+2.16	Strong warming
CH_4	Agriculture, FFs, landfills, wetlands	Indirect (ozone precursor)	~9 years	+0.54	Strong warming (GWP: 28–34)
N_2O	Agriculture, FFs, industrial processes	Indirect (stratospheric ozone depletion)	~121 years	+0.21	Warming (GWP: 265–298)
$PM_{2.5}$	Combustion, secondary formation	Cardiovascular, respiratory disease, mortality	Days-weeks	–0.27 to +0.20	Mixed
PM_{10}	Dust, combustion, industrial sources	Respiratory irritation, asthma exacerbation	Days-weeks	Variable	Generally cooling
BC	Incomplete combustion, BB	Cardiovascular, respiratory disease	Days-weeks	+0.40	Strong warming
Tropospheric O_3	Photochemical formation (No_x + VOCs)	Respiratory disease, crop damage	Weeks-months	+0.35	Warming
SO_2/ Sulfates	Coal combustion, industrial processes	Respiratory disease, acid rain	Days-weeks	–0.40	Strong cooling

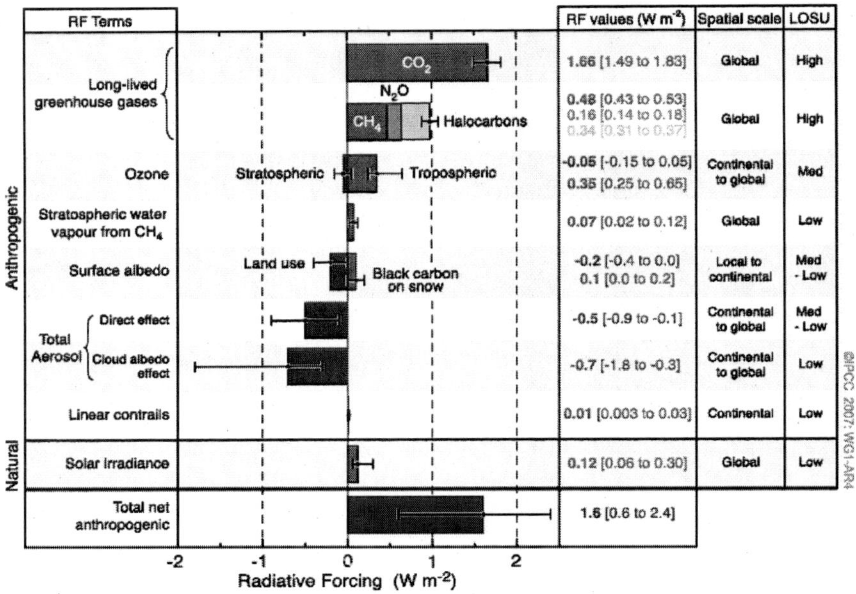

Figure 2: Global Average Radiative Forcing (RF) in 2005 Relative to 1750 for Major Climate Agents

Source: IPCC (2007), Climate Change 2007: The Physical Science Basis. Contribution of Working Group I to the Fourth Assessment Report of the Intergovernmental Panel on Climate Change, Figure SPM.2 (originally Figure 2.4).

III. Material Innovation for Atmospheric Pollution Control

Air pollution remains one of the most critical environmental challenges, with NO_x, SO_2, VOCs, and fine aerosols contributing to smog, acidic rain, and respiratory diseases. Recent advances in material science have yielded innovative solutions through photocatalytic materials, porous frameworks, and nanocomposites. Photocatalytic degradation relies on semiconductor materials such as Titanium dioxide (TiO_2), Zinc oxide (ZnO), or Graphitic carbon nitrite (g-C_3N_4) under light irradiation as illustrated in Figure 3.

TiO_2 remains the standard photocatalytic material with anatase and rutile phases. Several methods have been adopted to enhance the photocatalytic performance of TiO_2. In this regard, silver-doped TiO_2 reduces the band gap to about 2.85 eV, enhancing visible-light absorption and NO_x removal [29]. Another study developed a TiO_2-MXene composite that degraded 95% of tetracycline within 80 minutes under visible light through efficient charge separation [30]. Field tests in Denmark showed TiO_2-containing paving stones significantly reduced NO_x pollution outdoors [31]. Recent studies include anatase-rutile/BiOI composite fibers with enhanced visible light absorption [32], and ternary anatase-brookite-rutile N-doped TiO_2 showing superior methylene blue degradation [33]. Doping with noble metals (Ag, Pt, Au) effectively enhances visible light activity by reducing electron-hole recombination, with Ag doping particularly effective for generating reactive oxygen species.

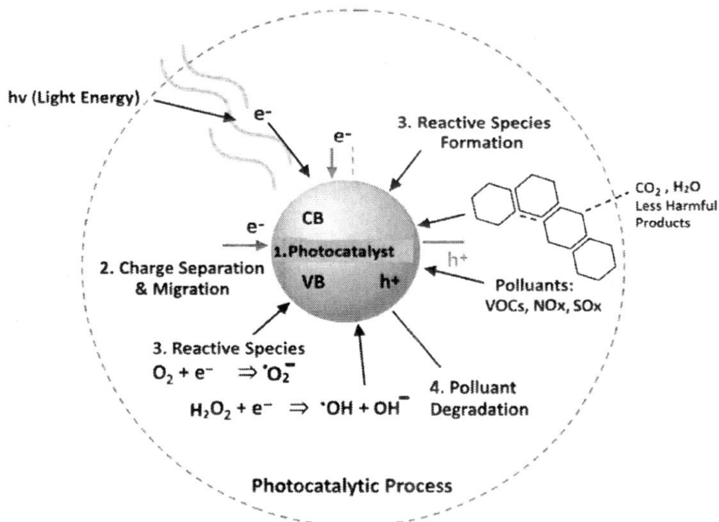

Figure 3: Mechanism of Photocatalytic Pollutant Degradation

Furthermore, metal-organic frameworks (MOFs) and covalent organic frameworks (COFs) have emerged as next-generation materials for gas capture. Functionalized MOF-74 variants achieve CO_2 uptakes exceeding 12 $mmol.g^{-1}$ under ambient conditions (Yang *et al.*, 2022). Amino-functionalized MIL-101 demonstrates selective CO_2 capture under humid flue gas [34], while copper-based MOFs reach SO_2 capacities above 15 $mmol.g^{-1}$ [35]. Nitrogen-rich COFs like TpPa-1 show superior CO_2 selectivity [36], and ionic-liquid modified COFs enhance CH_4 storage. Table 2 highlights MOFs superiority due to exceptionally high surface area, tunable porosity, and functionalization capability. COFs provide structural flexibility for selective adsorption, while zeolites remain effective but limited by fixed pore structures.

Besides, Carbon Nano Tubes (CNTs) and graphene derivatives offer high surface area and synergistic interactions with metal catalysts. MnO_2-CNT composites demonstrated 70% Nitric oxide (NO) reduction under real-exhaust conditions [37], while reduced graphene oxide (rGO)-TiO_2 systems achieved near-complete toluene degradation under visible light [38]. Ni-CNT hybrids offer excellent resistance to sintering in harsh environments, while Pt-graphene hybrids provide superior activity with reduced noble-metal loading [39]. Polymer-CNT membranes achieved more than 90% $PM_{2.5}$ capture efficiency [40]. Multifunctional composites combining CNTs with transition metal oxides both trap and catalytically oxidize BB generated aerosols [41], indicating versatility in environmental remediation.

Table 2: Comparative efficiency of MOFs, COFs, and zeolites for pollutant capture

Framework Type	*CO_2 Capture Efficiency*	*Methane Storage*	*Functionalization for Selective Adsorption*
MOFs	Very high uptake due to tunable pore sizes, high surface area (>3000 m^2/g), and open metal sites enhancing CO_2 affinity.	High volumetric and gravimetric storage capacity; widely studied for natural gas storage.	Easily functionalized with amine groups, open metal sites, or heteroatoms to enhance selectivity for CO_2, SO_2, NOx, VOCs.
COFs	Moderate to high capture; lighter frameworks with tunable pore size and chemical design.	Promising methane storage but generally lower than MOFs due to lower density.	High flexibility for chemical functionalization; tailored linkers allow selective adsorption of specific pollutants.
Zeolites	Good CO_2 capture but limited by fixed pore size and lower surface area compared to MOFs/COFs.	Moderate methane storage; restricted by pore geometry.	Functionalization possible through ion-exchange or surface modification, but less versatile than MOFs/COFs.

IV. Sensor Technologies and Monitoring

The air quality monitoring has evolved significantly with the advent of advanced sensor technologies. Traditional regulatory monitoring stations provide highly accurate data, but their widespread deployment is restricted by high costs, limited spatial coverage, and demanding maintenance requirements. In contrast, low-cost air quality sensors enable high density monitoring networks, that capture fine scale spatial variations in air pollution that were previously overlooked. These sensors are effective in identifying temporal changes, pollution hotspots, and providing real time alerts to support health protection. When integrated with air pollution control systems, such sensors facilitate closed loop feedback mechanisms that enhance pollutant removal efficiency. However, despite many advantages, there are some challenges such as sensor drift, cross-sensitivities, and regular calibration requirements.

A. Nanomaterial-Based Gas Sensors

Nanomaterial-based gas sensors have attracted significant attention due to their enhanced surface and electronic properties. ZnO nanowires show enhanced sensitivity toward CO and NO_2 gases due to their high surface to volume ratio and excellent electron mobility. Due to their unique electronic properties, CNTs based sensors exhibit exceptional sensitivity to trace gas concentrations through conductivity changes upon adsorption of gas. Hybrid sensors combining CNTs with metal oxides further improve performance by integrating high surface area and chemical reactivity.

Table 3: Performance Characteristics and Comparative Analysis of Advanced Gas Sensing Technologies for Environmental Monitoring Applications

Sensor Type	*Target Gas*	*Sensitivity (ppm)*	*Detection Limit (ppb)*	*Operating Temp (°C)*	*Response Time (s)*	*Stability*
ZnO Nanowires	NO_2	10–100	50–500	300–350	30–60	Good (>6 months)
ZnO Nanowires	CO	5–50	100–1000	300–350	60–120	Good (>6 months)
Au-ZnO Nanowires	NO_2	1–50	10–100	25–150	15–45	Excellent (>12 months)
CNT-ZnO Hybrid	NO_2	0.1–10	5–50	25–100	10–30	Excellent (>12 months)
CNT-based	NH_3	0.5–25	20–200	25–100	20–60	Good (>8 months)
SnO_2-ZnO Core Shell	CO	1–100	50–500	200–300	30–90	Good (>6 months)
Electrochemical	NO_2	0.01-10	1-20	25	30-120	Fair (3-6 months)
NDIR (Optional)	CO_2	10-10,000	1000-5000	25	10-60	Excellent (>24 months)
Metal Oxide (Commercial)	O_3	0.1-1	10-50	25-50	60-180	Fair (6-12 months)

B. Remote Sensing Technologies and Ground Validation

Satellite remote sensing provides global and continuous observations of atmospheric pollutants, addressing spatial gaps in ground monitoring networks. Onboard satellite sensors such as Tropospheric Monitoring Instrument (TROPOMI), Moderate Resolution Imaging Spectroradiometer (MODIS), and Visible Infrared Imaging Radiometer Suite (VIIRS) measure column concentrations of NO_2, SO_2, CO, and aerosols with high spatial and temporal resolution. Aerosol Optical Depth (AOD) indicates particulate loading, where values below 0.1 represent clean air and above 3.0 indicate severe pollution. Data reliability is ensured through ground validation campaigns by EPA, NASA, and NOAA, using surface and aircraft measurements. Differential Optical Absorption Spectroscopy (DOAS) and MAX-DOAS provide high resolution reference data for calibration and bias correction. The synergy of satellite observations and ground validation yields robust datasets for air quality modeling, forecasting, and regulatory applications.

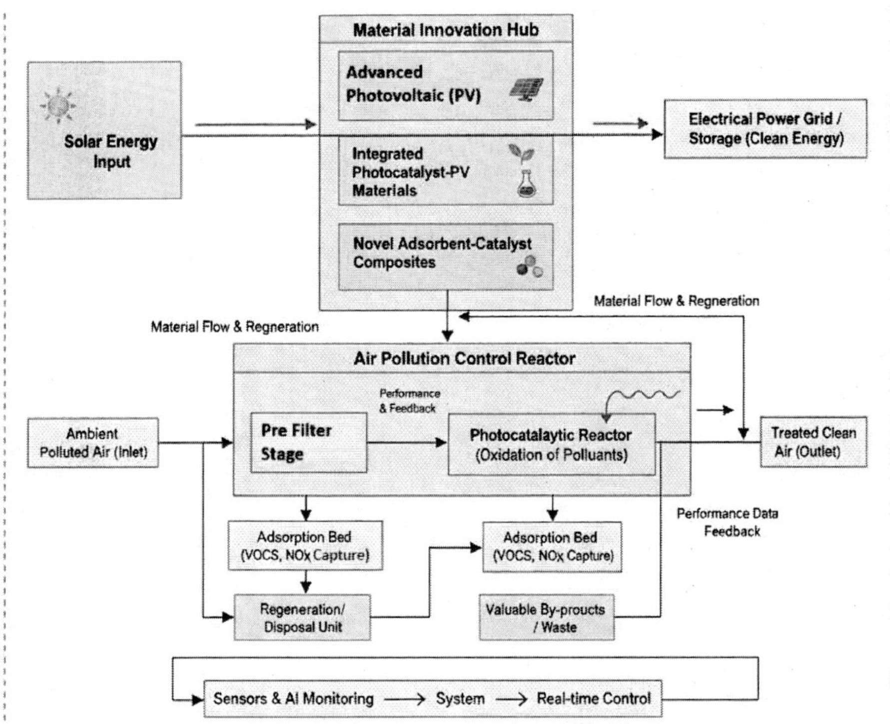

Figure 4: Integrated Material Innovation Framework for Solar-Driven Air Pollution Control

IV. Solar Energy Integration for Pollution Control

Integrating solar energy with pollution control technologies paves way for reduction in greenhouse gas emissions through renewable power generation while also enabling active pollutant removal. Using the cutting-edge catalysts like copper nanostructures, MOFs, and single atom systems, the solar driven electrochemical CO_2 reduction transforms atmospheric carbon into valuable fuels and chemicals. Over the past ten years, photovoltaic costs have decreased by around 90%, and copper-silver alloys have demonstrated remarkable ethylene selectivity. Tandem photovoltaic electrochemical cells have reached more than 20% solar-to-fuel efficiency. Concurrent developments in decentralized solar-powered air purification incorporate HEPA, activated carbon, and TiO_2-based photocatalytic modules, achieving 75–85% degradation of VOCs and NO_x. These autonomous systems, costing $50–200 m^{-2} have a lifespan of 15–25 year and 2–4-year carbon payback. They are sustainable clean-air solutions powered solely by renewable energy.

VI. Methodology, Research Approaches and Societal Relevance

A systematic evaluation of materials for atmospheric sustainability requires an integrated methodology that combines experimental synthesis, computational modeling, field validation, and life-cycle evaluation, shown in Table 4. Such comprehensive framework guarantees the effective translation of laboratory scale innovations to practical pollution control technologies. Material synthesis is achieved using sol-gel, hydrothermal, vapor-phase, and electrospinning techniques, while the characterization is conducted through X-ray diffraction, surface area analysis, electron microscopy, and spectroscopic methods. Adsorption and photocatalytic testing techniques measure the efficiencies to capture pollutants and their degradation under controlled conditions. Computational tools such as density functional theory, molecular dynamics, and Monte Carlo simulations provide insight into electronic and adsorption mechanisms. Machine learning models enhance the efficiency of material discovery and predict the performance. Field validation through urban and industrial testbeds assesses durability and operational efficiency, while life-cycle evaluation examines resource utilization, emissions, and cost-benefit trade-offs to ensure sustainable implementation.

Table 4: Overview of Methodologies used in Material Evaluation for Pollution Control

Methodology Category	*Specific Techniques*	*Key Outputs*	*Timeline*	*Resource Requirements*
Experimental Synthesis	Sol-gel, hydrothermal, CVD, electrospinning	Nanomaterials with controlled properties	1–6 months	Specialized equipment, precursor materials
Material Characterization	XRD, SEM/TEM, BET, UV-Vis, FTIR	Crystal structure, morphology, surface area, optical properties	2–4 weeks	Analytical instruments, trained personnel
Adsorption Testing	Breakthrough curve analysis, isothermal experiments	Capture capacity, selectivity, kinetics	2–8 weeks	Gas delivery systems, controlled chambers
Photocatalytic Assays	Laboratory photoreactors, quantum efficiency measurement	Degradation rates, reaction mechanisms	2–6 weeks	Light sources, analytical monitoring
DFT Calculations	Electronic structure modeling, band structure analysis	Band gaps, charge transport properties	1–4 weeks	High-performance computing, software licenses
Molecular Simulations	GCMC, molecular dynamics	Adsorption isotherms, binding energies	2–6 weeks	Computational resources, validated force fields
Field Validation	Urban testbeds, industrial pilots	Real-world performance, durability assessment	6–24 months	Installation costs, monitoring equipment
Life-cycle Assessment	ISO 14040/14044 standards, impact quantification	Environmental footprint, cost-benefit analysis	2–6 months	LCA software, comprehensive data inventory

Adhering to the United Nations Sustainable Development Goals, particularly SDG 3 (Good Health and Well-being), SDG 7 (Affordable and Clean Energy), SDG 11 (Sustainable Cities and Communities), and SDG 13 (Climate Action), these research methodologies transform laboratory findings into practical solutions. Photocatalytic coatings, MOF-based filters, and nanostructured adsorbents represent solutions that concurrently diminish pollution exposure and enhance climate resilience. Innovations integrated with policy frameworks and economic life-cycle analyses promote sustainable development pathways; wherein sophisticated materials serve as essential facilitators of cleaner air, renewable energy integration, and enduring environmental stability.

VII. CONCLUSION

The present chapter highlights the role of materials innovation in delivering integrated solutions for mitigating air pollution and climate change. It has been discussed that the advanced materials such as photocatalytic nanostructures, carbon nanocomposites, porous frameworks, and smart sensors, enable pollutant removal, renewable energy conversion, and real time atmospheric monitoring. Quantitatively, metal organic frameworks capture 15–20 wt% CO_2, photocatalysts achieve 80–95% pollutant degradation, and urban applications reduce $PM_{2.5}$ by 20–40%, at estimated cost of $50–150 per ton of CO_2-equivalent removed. Achieving large scale impact requires multidisciplinary collaboration across materials science, environmental engineering, and policy framework. As the global temperature surpasses 1.5°C and 99% of humanity breathes polluted air, advancement in material technologies is crucial for achieving sustainable air quality and climate stabilization, delivering equal benefits for health, energy, and the environment.

REFERENCES

[1] WHO (World Health Organization), WHO Global Air Quality Guidelines: Particulate matter (PM2.5 and PM10), ozone, nitrogen dioxide, sulfur dioxide and carbon monoxide, World Health Organization (2021).

[2] I. Manisalidis, E. Stavropoulou, A. Stavropoulos and E. Bezirtzoglou, Environmental and health impacts of air pollution: A review, Front. Public Health 8, 14 (2020).

[3] R. Fuller, P.J. Landrigan, K. Balakrishnan, G. Bathan, S. Bose-O'Reilly, M. Brauer, *et al.*, Pollution and health: A progress update, Lancet Planet. Health 6, e535 (2022).

[4] J. Lelieveld, K. Klingmüller, A. Pozzer, R.T. Burnett, A. Haines and V. Ramanathan, Effects of fossil fuel and total anthropogenic emission removal on public health and climate, Proc. Natl. Acad. Sci. 116, 7192 (2020).

[5] IPCC (Intergovernmental Panel on Climate Change), Climate Change 2023: Synthesis Report. Contribution of Working Groups I, II and III to the Sixth Assessment Report, Geneva, Switzerland: IPCC (2023).

[6] P.M. Forster, C. Smith, T. Walsh, W.F. Lamb, R. Lamboll, B. Hall, *et al.*, Indicators of Global Climate Change 2022: Annual update of large-scale

indicators of the state of the climate system and human influence, Earth Syst. Sci. Data 15, 2295 (2023).

[7] D. Shindell, J.C. Kuylenstierna, E. Vignati, R. van Dingenen, M. Amann, Z. Klimont, *et al.*, Simultaneously mitigating near-term climate change and improving human health and food security, Science 335, 183 (2012).

[8] E. Von Schneidemesser, P.S. Monks, J.D. Allan, L. Bruhwiler, P. Forster, D. Fowler, *et al.*, Chemistry and the linkages between air quality and climate change, Chem. Rev. 115, 3856 (2015).

[9] X. Lan, P. Tans, K.W. Thoning and NOAA Global Monitoring Laboratory, Trends in globally-averaged CO_2 determined from NOAA Global Monitoring Laboratory measurements, NOAA/GML (2023).

[10] T.C. Bond, S.J. Doherty, D.W. Fahey, P.M. Forster, T. Berntsen, B.J. DeAngelo, *et al.*, Bounding the role of black carbon in the climate system: A scientific assessment, J. Geophys. Res.: Atmos. 118, 5380 (2013).

[11] B.H. Samset, M.T. Lund, M. Bollasina, G. Myhre and L. Wilcox, Emerging Asian aerosol patterns, Nat. Geosci. 12, 582 (2022).

[12] N. Bellouin, J. Quaas, E. Gryspeerdt, S. Kinne, P. Stier, D. Watson-Parris, *et al.*, Bounding global aerosol radiative forcing of climate change, Rev. Geophys. 58, e2019RG000660 (2020).

[13] A. Markandya, J. Sampedro, S.J. Smith, R. Van Dingenen, C. Pizarro-Irizar, I. Arto and M. González-Eguino, Health co-benefits from air pollution and mitigation costs of the Paris Agreement: A modelling study, Lancet Planet. Health 2, e126 (2018).

[14] N. Scovronick, M. Budolfson, F. Dennig, M. Fleurbaey, A. Siebert, R.H. Socolow, *et al.*, Impact of population growth and population ethics on climate change mitigation policy, Proc. Natl. Acad. Sci. 114, 12338 (2019).

[15] Y. Zhang, J.J. West, R. Mathur, J. Xing, C. Hogrefe, S.J. Roselle, *et al.*, Long-term trends in the ambient $PM_{2.5}$ and O_3-related mortality burdens in the United States under emission reductions from 1990 to 2010, Atmos. Chem. Phys. 18, 15003 (2022).

[16] T. Vandyck, K. Keramidas, S. Tchung-Ming, M. Weitzel and R. Van Dingenen, Quantifying air quality co-benefits of climate policy across sectors and regions, Clim. Change 163, 1365 (2023).

[17] D. Chen, Y. Cheng, N. Zhou, P. Chen, Y. Wang, K. Li, *et al.*, Photocatalytic degradation of organic pollutants using TiO_2-based photocatalysts: A review, J. Clean. Prod. 268, 121725 (2021).

[18] R.S. Haszeldine, S. Flude, G. Johnson and V. Scott, Negative emissions technologies and carbon capture and storage to achieve the Paris Agreement commitments, Philos. Trans. R. Soc. A 376, 20160447 (2018).

[19] S. Kumar, J.G. Sharma, S. Maji and B.D. Malhotra, Nanostructured zirconia decorated reduced graphene oxide based efficient biosensing platform for non-invasive oral cancer detection, Biosens. Bioelectron. 78, 497 (2020).

[20] N. Afshar-Mohajer, C. Lam, M.C. Hatzell and A. Laborte, Nanomaterial-based sensors for air quality monitoring: Recent advances and future perspectives, Environ. Sci.: Nano 10, 645 (2023).

[21] E. Boonen and A. Beeldens, Recent photocatalytic applications for air purification in Belgium, Coatings 4, 553 (2014).

[22] D. Chakraborty, K. Gupta and S. Basu, Nanomaterial-based gas sensors for air quality monitoring: Challenges and opportunities, Sens. Actuators B: Chem. 355, 131338 (2022).

[23] J.H. Seinfeld and S.N. Pandis, Atmospheric Chemistry and Physics: From Air Pollution to Climate Change (3rd ed.), John Wiley & Sons (2016).

[24] O. Boucher, D. Randall, P. Artaxo, C. Bretherton, G. Feingold, P. Forster, *et al.*, Clouds and aerosols, in Climate Change 2013: The Physical Science Basis. Contribution of Working Group I to the Fifth Assessment Report of the Intergovernmental Panel on Climate Change, Cambridge University Press, pp. 571-657 (2013).

[25] EPA (U.S. Environmental Protection Agency), Climate Change Indicators: Atmospheric Concentrations of Greenhouse Gases, Retrieved from https://www.epa.gov/climate-indicators/ (2025).

[26] PMC (PubMed Central), Black Carbon: A Review of Current Knowledge and Policy Recommendations, Environ. Health Perspect. 130 (2022).

[27] MIT Climate Portal, Climate Change Basics: Radiative Forcing, Massachusetts Institute of Technology, Retrieved from https://climate.mit.edu/.

[28] IPCC, 2007: Climate Change 2007: The Physical Science Basis. Contribution of Working Group I to the Fourth Assessment Report of the Intergovernmental Panel on Climate Change. Cambridge University Press, Cambridge, United Kingdom and New York, NY, USA.

[29] H. Si, Photocatalytic concrete for NO_x degradation: influence of TiO_2 dispersion in cement mortar, Mater. Sci. (2021).

[30] S.M. Mousavi, M.S. Mohtaram, K. Rasouli, S. Mohtaram, H. Rajabi and S. Sabbaghi, Efficient visible-light-driven photocatalytic degradation of antibiotics in water by MXene-derived TiO_2-supported SiO_2/Ti_3C_2 composites: Optimisation, mechanism and toxicity evaluation, Environmental Pollution, 367, 125624 (2025). DOI: 10.1016/j.envpol.2024.125624

[31] A. Folli, *et al.*, Field study of air purifying paving elements containing TiO_2, Sci. Total Environ. (2015).

[32] D. Li, K. Xu and C. Zhang, Improvement of photocatalytic performance by building multiple heterojunction structures of anatase–rutile/BiOI composite fibers, Nanomaterials 12, 3906 (2022).

[33] E. Assayehegn, A. Solaiappan, *et al.*, In situ–driven formation of anatase/brookite/rutile heterojunction N/TiO_2 nanocrystals as sustainable visible-light catalysts, Global Challenges 8, 2400174 (2024).

[34] J. Zheng, *et al.*, Amino-functionalized MIL-101 for CO_2 capture under humid conditions, Chem. Eng. J. (2021).

[35] Z. Wang, *et al.*, SO_2 capture in Cu-based MOFs, Chem. Rev. (2020).
[36] T. Zhou, *et al.*, Nitrogen-rich COFs for CO_2 capture and selectivity, Angew. Chem. (2019).
[37] Y. Gao, *et al.*, MnO_2-CNT nanocomposites for NO reduction in automotive exhaust, Catal. Today (2020).
[38] H. Kim, *et al.*, Reduced graphene oxide-TiO_2 composites for VOC degradation, J. Hazard. Mater. (2021).
[39] L. Huang, *et al.*, Pt-graphene hybrids in catalytic converters: durability and cost reduction, ACS Appl. Mater. Interfaces (2022).
[40] M. Zhu, *et al.*, CNT-polymer hybrid membranes for $PM_{2.5}$ capture, J. Membr. Sci. (2022).
[41] R. Singh, *et al.*, Hybrid nanocomposites for biomass aerosol capture and oxidation, Environ. Pollut (2023).

CHAPTER 11

A Review on Thermochemical Treatment of Biomass and Plastics for Sustainable Hydrogen Production

Shekhar Jyoti Pathak

Department of Chemical Engineering,
Indian Institute of Technology Guwahati, Assam, India
E-mail: shekharjyotipathak@gmail.com

ABSTRACT: Global energy sector mainly depends on the traditional fossil fuels like crude oil and coal. Their extensive utilization leads to global warming, climate change, environmental and human health issues as well as resource depletion. Clean alternative energy sources are essential to prevent the environmental damages and strengthen the energy security. In this context, hydrogen is a well-recognized renewable fuel providing carbon neutrality and efficient energy solutions. Considering the growing requirement in various fields from industry to transport, hydrogen has a major role to play in transitioning the global energy sector towards sustainability by the year 2050. Affordability and abundance make biomass a significant source for renewable energy production. On the other hand, enormous use of non-biodegradable plastics have augmented a growing interest for effective waste management to prevent its harmful effect on the environment. Biomass and waste plastics can serve as potential alternative sources to fossil fuels for high value utilization in the energy sector. Considering the global importance of clean hydrogen fuel production, this chapter intends to provide a review on the thermochemical conversion technology for biomass and plastics. Two popular methods known as gasification and pyrolysis are discussed broadly. Challenges and scope in thermochemical treatment of biomass and plastics during hydrogen production are also highlighted.

Keywords: Energy, Hydrogen, Thermochemical Process, Gasification, Pyrolysis.

I. INTRODUCTION

Hydrogen (H_2) is a clean energy carrier and potential contributor in industrial and environmental sustainability. Being a non-hazardous energy career, it ensures safety during storage, utilization and transportation. To fulfil the escalating demand, global energy consumption is estimated to increase by 70.00% in the period 2018–2050 [1]. In this context, annual demand and supply for H_2 is also rising and it will reach around 10 Exajoul (EJ) by 2050, with approximately 5.00–10.00% annual rise after 2050 [2]. H_2 production is around 50 million metric tons (MMT)/year and currently, it is used in the ammonia production (51.00%), oil refining (31.00%), methanol production (10.00%) and for other needs (8.00%) [2, 3]. H_2 is produced as green, blue or brown hydrogen through fossil fuel reforming, water electrolysis, thermolysis or photolysis. As shown in Figure 1, H_2 utilization in variuos sectors makes it ideal for achieving low-carbon energy future.

Biomass is a renewable and sustainable energy source and more promising for H_2 production compared to the fossil fuels. On the other hand, rapid utilization and inappropriate treatment of the plastic made equipment have uplifted the global plastic generation from 270 MMT in 2000 to 445 MMT in 2025 and further it is going to reach around 590 MMT by 2050 [4]. Electronic waste (EW) is a major contributor of non-biodegradable and harmful waste plastics as 20.00-35.00% of EW is nothing but plastic materials [5]. Incineration and landfilling of these plastics with poor disposal strategies have caused major environmental deterioration and human health issues. Conversion of the waste plastics into H_2 fuel through thermochemical recycling can help in achieving carbon neutrality and improves the waste management. Considering abundantly available biomass and the staggering issues of waste plastic management, thermochemical treatment can lead to a sustainable way of converting them into energy products. Because the gaseous components obtained in the thermochemical processing of biomass

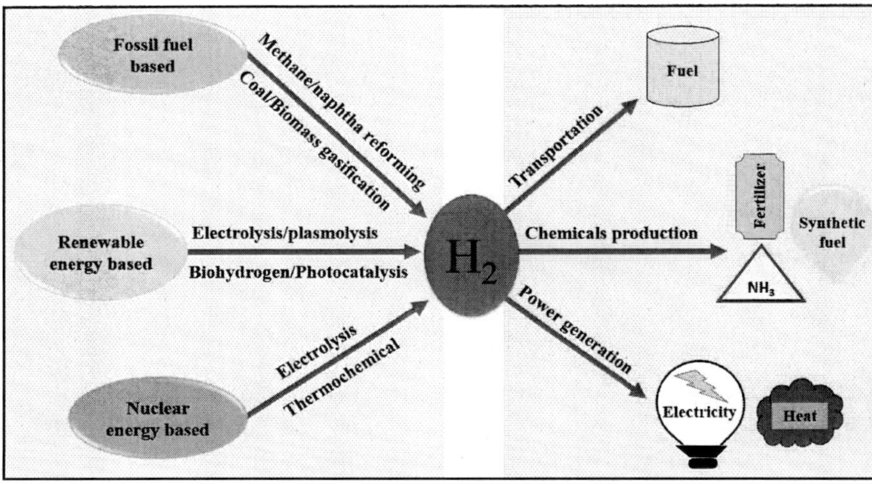

Figure 1: Hydrogen Production and Utilization

and waste plastics are effective for production of H_2 and carbon materials [6]. This chapter mainly deals with the H_2 fuel production from renewable biomass and plastics via thermochemical treatment namely, gasification and pyrolysis. Additionally, biomass and plastics co-processing through these advanced techniques are also discussed.

II. GASIFICATION FOR H_2 PRODUCTION

Gasification is a popular thermochemical route to convert carbonaceous materials into gaseous products at operating temperature 600–900°C using gasifying agent [7, 8]. It converts biomass and plastics into a fuel gas called syngas through a series of homogeneous and heterogeneous reactions which is mainly comprising of H_2, CO, CO_2, C_nH_m [7]. The resulting syngas can be used in various applications including energy generation, fuels and chemicals production.

A. Biomass Gasification

Lignocellulosic biomass and various organic wastes can be successfully gasified to produce synthesis gas (CO and H_2) used for power generation, chemicals production followed by CO to pure H_2 conversion that is used in fuel cells [9, 10]. Gasification of biomass usually takes place through a sequence of complex thermochemical reactions representing different stages of drying, pyrolysis, combustion and reduction zones [11]. Usually, air, pure oxygen, steam, carbon dioxide, nitrogen or mixtures act as the gasifying agent in biomass gasification. In steam gasification (SG), steam enhances H_2 formation and generates gases with high heating value in absence of nitrogen [2]. Researchers performed SG using the fluidized bed reactor and reported significant gas production (10–16 MJ/Nm^3) containing 30.00–60.00% of H_2 [11]. Catalytic SG of pine sawdust resulted in syngas with attractive H_2 content beyond 50.00 vol% [12]. Also, high temperature SG favoured woody biomass conversion to hydrogen-rich syngas (35.00–55.00 vol% of H_2) effectively [13]. Besides, supercritical water gasification (SWG) is an efficient and clean conversion pathway for biomass. Nearly, complete conversion of the feedstock (corn silage and clover grass) was achieved at temperatures beyond 700°C with H_2 content 26.00%–57.00% [14]. On the other hand, chemical looping gasification (CLG) with an oxygen carrier is a low cost technology to convert biomass into H_2 rich syngas. Agricultural biomass gasification in dual fluidized-bed reactors using steam (gasifying agent) by CLG resulted in a producer gas with 71.00% of H_2 concentration and nearly 0% CO_2 existence [15].

B. Plastic Gasification

In gasification operation, plastics usually react with gasifying agent (steam, oxygen or air) at high temperature 500°C–1300°C and generate valuable hydrocarbons, particularly H_2 rich syngas [4,16]. Plastic to syngas conversion includes pyrolysis, reforming, and secondary gasification, where, thermally cracked intermediate hydrocarbons from the pyrolysis treatment get reformed to H_2 and CO [8]. Combustion and partial oxidation play a major role in plastics to gas conversion during oxygen or air gasification. On the other hand, SG of plastics usually gives syngas with rich hydrogen content and a high H_2/CO ratio [17]. It was reported that more than 60.00 vol% of H_2 in syngas can be obtained via steam reforming reaction of high density polyethylene (HDPE) pyrolysis gas in a two-stage gasifier [10]. Further, catalyst can accelerate the plastic gasification efficiency by maximizing H_2 content in the resulting gaseous mixture and lowering the hydrocarbon and methane content [18]. During catalytic gasification, volatiles and tar come in contact with the catalyst bed in presence of CO_2 flow and decompose quickly to produced more H_2 [19]. Researchers performed polyethylene gasification in a fixed bed reactor using Ni–Al_2O_3 catalyst and reported the gas yield upto 2.04 m^3/kg at 900°C with H_2 concentration around 37.00 wt% [20].

III. Pyrolysis for H_2 Production

Pyrolysis, unlike gasification works under lower temperatures and without oxidizing agent. In absence of air and oxygen, typically, 400–900°C temperature and 1–5 bar of pressures are considered for pyrolysis experiments [2,21]. Pyrolytic products are gases (mainly H_2, CO, CO_2, CH_4 and H_2O), liquids (bio-oil) and solids (char). Such technologies can play major role in clean energy production by reducing the emission of COx with a healthy solid carbon recovery.

A. Biomass Pyrolysis

Concentration of H_2 in syngas generally varies from 1.50 to 45.00 vol% in fast pyrolysis and from 9.00 to 44.00 vol% in flash pyrolysis [22]. In slow pyrolysis of woody biomass, around 42.00 vol% of H_2 was found in resulting syngas [21]. To enhance direct H_2 production, biomass flash pyrolysis in series with steam reforming is considered effective. During this operation, unquenched bio-oil vapors from one reactor are subjected to the reforming reactions in another reactor with catalyst addition for higher H_2 yield [23]. Pyrolytic bio oil obtained from biomass was reformed using novel metal-doped catalysts. Highest yield of H_2 and carbon conversion were reported as 80.00% and 96.00%; respectively [24]. In cellulose pyrolysis integrated with catalytic plasma reforming, H_2 production was maximized to 26.60 mmol/g of H_2 at 550°C using Ni-Co/Al_2O_3 catalyst [25]. Whereas, steam pyrolysis of olive pomace at 500–700°C showed highest H_2 yield around 14.07 mmol/g [26]. Besides, microwave assisted catalytic pyrolysis of water hyacinth biomass was carried out in a two-stage reactor, where, highest H_2 yield of 77.20% was recorded using a Ni-based catalyst at 800°C and 17 min of residence time [27]. These integrated catalytic pyrolysis operations favor product selectivity with improved H_2 concentration in the syngas. However, temperature, residence time, process circumstances, catalysts and feedstock properties play vital role in enhancing biomass to H_2 conversion [28].

B. Plastic Pyrolysis

Usually, thermal pyrolysis of plastics requires a temperature range 400–600°C and for better H_2 yield, catalytic steam reforming is preferred as a post-treatment. According to this concept, plastic is liquefied (oil) by thermal decomposition, which undergoes steam reforming to produce synthetic gas with high H_2 concentration [29]. The pyrolysis/reforming operation of polypropylene (PP) resulted in H_2 production, approximately 170 mmol/g of PP [30]. This combined strategy was also applied for PS plastics and recorded 62.26 mmol H_2/g of PS in presence of Ni/Al_2O_3 catalyst [31]. Another study using polyolefin showed H_2 yields around 130 mmol/g plastic, whereas, industrial and commercial mixed plastics produced H_2 in a range 92.81 to 122.60 mmol/g of plastics [32]. Microwave pyrolysis facilitates rapid heating, short reaction time, and low energy loss and a good alternative to the conventional pyrolysis having drawbacks like high heat loss and low heating efficiency [33]. Microwave pyrolysis of polyethylene (PE) gave H_2 concentration up to 67.85 vol% and yield of 48.10 mmol/g plastic; respectively [34]. Significant enhancement in H_2 yield (from 25.00 to 50.20 mmol/g) and its selectivity (from 66.00 to 91.50 vol%) were also reported in microwave pyrolysis of HDPE compared to conventional thermal pyrolysis [35]. Research studies are evidence of pyrolysis being potential contributor in global waste management through waste plastics recycling and also offering fuel production.

IV. Co-processing of Biomass and Plastics

Single plastic gasification has some major problems like material softening/stickiness and coke formation, which make the operation complex. Besides, biomass gasification also faces low H_2/CO ratio in syngas with higher amount of unwanted tar/char formation. Thus, co-processing plastics with biomass through gasification is a possible way of minimizing the abovementioned issues encountered with individual feedstock. Thermochemical interaction between plastics and biomass leads to higher syngas gas yield and reduced tar formation [6, 36, 37]. For example, polyethylene (PE) addition was found benificial in reducing both tar and char during co-gasification. As per report, tar content significantly reduced to only 9.74 gN/m^3 during co-gasification compared to 58.23 gN/m^3 obtained in the single biomass gasification [38]. Addition of plastics in biomass gasification favoured H_2 (50.00 vol%) and reduced CO concentration in gaseous product [39].

Like gasification, co-processing biomass and plastics via pyrolysis shows synergistic chemical interaction between their thermally cracked intermediate species, affecting the desired product yield and quality [40]. In co-processing with biomass, increasing the plastic content from 30.00 to 50.00 wt.% enhanced the H_2 yield from 102.52 mmol/g to 115.33 mmol/g [18]. Pinewood and polypropylene (PP) co-pyrolysis also showed major improvements in syngas production and H_2 yield by 27.00% and 80.00%; respectively [41]. Therefore, thermochemical co-processing of plastics with biomass not only favors the H_2 production, but also provides a potential route for global plastic waste management (Table 2). A schematic of thermochemical treatment pathways for biomass/plastic mixture for H_2 production is shown in Figure 2.

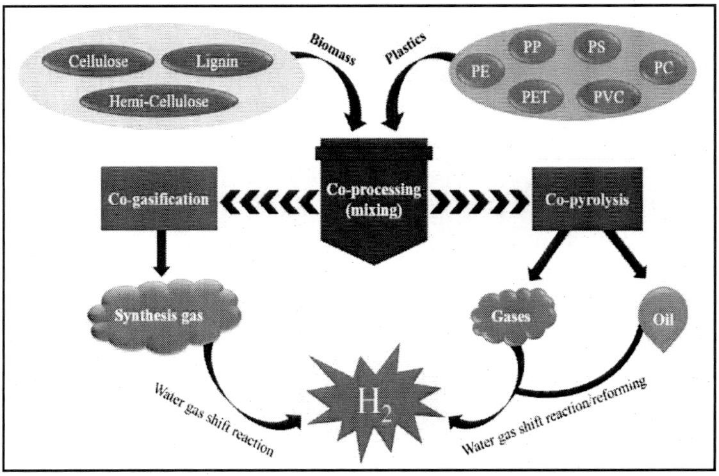

Figure 2: Schematic of Biomass and Plastics Co-processing via Thermochemical Routes for H_2 Production

Table 2: Co-processing of Biomass and Plastics for H_2 Production

Feedstock [Process]	*Operational Parameter*	*H_2 production*	*Remarks*	*References*
Rice husk (RH) and PE [Co-gasification]	850°C 80.00% RH, 20.00% PE	≈ 41.00 vol%	Steam promoted H_2, air increased CO and CO_2 in syngas	[42]
Rice straw (RS), PVC and PP [Co-gasification]	900°C 20.00% RS, 40.00% PVC, 40.00% PP	49.10 vol%	Volatile–char interaction and mineral catalytic effects might have occurred due to blending	[36]
Pine cones (PC) and PP [Co-gasification]	700°C 30.00% PC, 70.00% PP	52.60 vol%	Water gas shift reaction dictated the final gas composition	[43]
Pine wood (PW) and HDPE [Co-gasification]	900°C 50.00% PW, 50.00% HDPE	57.00%	Co-gasification enhanced the carbon conversion efficiency and favoured gas yield	[38]
Coconut shell (CS) and PE [Co-gasification]	800°C 50.00% CS, 50.00% PE	81.60 vol%	High carbon conversion efficiency (92.40%) and H_2 rich syngas are impressive for scale up	[44]
Paper biomass (PB) and plastics (P) [Co-pyrolysis]	800°C, 15 min PB/P ratio 5:1 Co-CeO_2/Al_2O_3 catalyst	47.00 vol%	Catalytic performance on co-pyrolysis enhanced with increase in Co loading	[45]
Bamboo (B) and HDPE [Co-pyrolysis]	300–900°C 20.00% B, 80.00 % HDPE Ni/Al_2O_3 catalyst	54.00 vol%	Microwave assisted co-pyrolysis not only enhanced H_2 production, but also improved oil quality	[46]
Cellulose and HDPE [Co-pyrolysis]	850°C Ni/MCM-41 catalyst Steam/feed ratio =4.85	64.64 vol%	Cellulose/plastic mixture produced more H_2 compared with lignin/plastic mixture	[47]

Feedstock [Process]	*Operational Parameter*	*H_2 production*	*Remarks*	*References*
Douglas fir and LDPE [Co-pyrolysis]	650°C, 15 min Biochar based catalyst	72.00 vol%	Biochar has good reusability in co-pyrolysis for higher H_2 production	[48]
Pine Sawdust (PSD) and HDPE [Co-pyrolysis]	700°C 25.00% PSD, 75.00 % HDPE	72.00 vol%	H_2 production increased lineally with HDPE content in the feed	[49]

V. Challenges and Scope

Thermochemical technologies encounter several bottlenecks and challenges in processing biomass/plastics feedstock for fuel production. In biomass gasification, undesirable tar formation hinders the quick H_2 accessibility and causes fouling and plugging in reactor system [50]. High operating cost and energy requirement, reactor corrosion are often encountered during SCW biomass gasification [51]. Catalyst addition provides significant scope to reduce the temperature and pressure required in SCW gasification for enhanced H_2 production, thereby, suppresses the cost [11]. Another important issue lies with the seasonal availability of biomass that can restrict the reliable H_2 production. Similarly, waste plastics gasification also suffers several hurdles, mainly softening and stickiness with coke formation that complicate the operational procedure [52]. In this context, co-processing plastics with biomass has emerged as a crucial thermochemical route for enhanced H_2 production with improved heating value of the producer gas and provides a good scope for optimization of the process on industrial point of view [52, 53]. Microwave assisted pyrolysis also requires more research based investigation to restrict harmful aromatics formation, better reactor construction for operational stability, metal and sulphur removal from feedstock [54]. Realising that thermochemical techniques release CO_2, researchers get scope for broad analysis on carbon capture and storage to contribute to a low-carbon hydrogen economy [11]. Countries like India with a huge biomass resource and enormous generated waste plastics should pay attention towards thermochemical conversion of waste materials into fuels. This will provide a significant scope to address the country's renewable energy demand through clean energy production and establishing India as a clean energy producer rather than remaining as a global polluter.

VI. Conclusions

Biohydrogen emerges as a novel energy career which is renewable in nature, clean and environmental friendly. Thermochemical technologies are promising in upcycling biomass and plastics into valuable fuels including H_2. Co-thermal processing of biomass and plastics have shown synergistic effect in improving syngas quality and H_2 yield. Further research investigation on sustainability assessment, techno-economic analysis, process parameter optimization, feedstocks and reactor design selection, suitable catalyst development should be encouraged to overcome the challenges encountered in sustainable H_2 production in large industrial scale.

References

[1] S.J. Pathak, A. Gangal and V. Prabu. Direct liquefaction of discarded printer cartridge plastics and its Kinetic Modelling. Fuel Process. Technol. [Internet] 2022 [cited 2026 Jan 16];228:107147. Available from: https://linkinghub.elsevier.com/retrieve/pii/S0378382021004306

[2] E.B. Agyekum, C. Nutakor, A.M. Agwa and S. Kamel. A Critical Review of Renewable Hydrogen Production Methods: Factors Affecting Their Scale-Up and Its Role in Future Energy Generation. Membranes [Internet] 2022 [cited 2026 Jan 13];12:173. Available from: https://www.mdpi.com/2077-0375/12/2/173

[3] R. Yukesh Kannah, S. Kavitha, Preethi, O. Parthiba Karthikeyan, G. Kumar, Dai-Viet NVo, *et al.* Techno-economic assessment of various hydrogen production methods – A review. Bioresour. Technol. [Internet] 2021 [cited 2026 Jan 13];319:124175. Available from: https://linkinghub.elsevier.com/retrieve/pii/S0960852420314498

[4] M. Lahafdoozian, H. Khoshkroudmansouri, S.H. Zein and A.A. Jalil. Hydrogen production from plastic waste: A comprehensive simulation and machine learning study. Int. J. Hydrog. Energy [Internet] 2024 [cited 2026 Jan 13];59: 465–79. Available from: https://linkinghub.elsevier.com/retrieve/pii/S0360319924003628

[5] S.J. Pathak and P. Vairakannu. Co-liquefaction of high ash coal and electronic waste plastics: Study of liquefaction reaction kinetics and characterization of the co-liquefied oil. Fuel [Internet] 2025 [cited 2026 Jan 16];390:134754. Available from: https://linkinghub.elsevier.com/retrieve/pii/S0016236125004788

[6] Q. Shang, D. Feng, X. Zhang, Z. Cheng, S. Wang and Y. Zhao, *et al.* Synergistic effects of biomass/plastics and multi-step regulation of H_2O in the co-production of H2-CNTs by gasification of biomass and plastic wastes. Appl. Catal. B Environ. Energy [Internet] 2025 [cited 2026 Jan 13];379: 125695. Available from: https://linkinghub.elsevier.com/retrieve/pii/S0926337325006782

[7] F. Ardolino, C. Lodato, T.F. Astrup and U. Arena. Energy recovery from plastic and biomass waste by means of fluidized bed gasification: A life cycle inventory model. Energy [Internet] 2018 [cited 2026 Jan 13];165:299–314. Available from: https://linkinghub.elsevier.com/retrieve/pii/S0360544218319303

[8] M.A. Bashir, T. Ji, J. Weidman, Y. Soong, M. Gray, F. Shi, *et al.* Plastic waste gasification for low-carbon hydrogen production: a comprehensive review. Energy Adv. [Internet] 2025 [cited 2026 Jan 13];4:330–63. Available from: https://xlink.rsc.org/?DOI=D4YA00292J

[9] L. Cao, I.K.M. Yu, X. Xiong, D.C.W. Tsang, S. Zhang, J.H. Clark, *et al.* Biorenewable hydrogen production through biomass gasification: A review and future prospects. Environ. Res. [Internet] 2020 [cited 2026 Jan 13];186:109547. Available from: https://linkinghub.elsevier.com/retrieve/pii/S0013935120304400

[10] H. Jeong, M. Hauser, F. Fischer, M. Hauck, S. Lobe, R. Peters, *et al.* Utilization of Bio-Syngas in Solid Oxide Fuel Cell Stacks: Effect of Hydrocarbon Reforming. J. Electrochem. Soc. [Internet] 2019 [cited 2026 Jan 13];166: F137–43. Available from: https://iopscience.iop.org/article/10.1149/ 2.1191902jes

[11] B. Dou, H. Zhang, Y. Song, L. Zhao, B. Jiang, M. He, *et al.* Hydrogen production from the thermochemical conversion of biomass: issues and challenges. Sustain. Energy Fuels [Internet] 2019 [cited 2026 Jan 13];3:314–42. Available from: https://xlink.rsc.org/?DOI=C8SE00535D

[12] S. Luo, B. Xiao, X. Guo, Z. Hu, S. Liu and M. He. Hydrogen-rich gas from catalytic steam gasification of biomass in a fixed bed reactor: Influence of particle size on gasification performance. Int. J. Hydrog. Energy [Internet] 2009 [cited 2026 Jan 13];34:1260–4. Available from: https://linkinghub.elsevier.com/retrieve/pii/S0360319908014729

[13] K. Umeki, K. Yamamoto, T. Namioka and K. Yoshikawa. High temperature steam-only gasification of woody biomass. Appl. Energy [Internet] 2010 [cited 2026 Jan 13];87:791–8. Available from: https://linkinghub.elsevier.com/retrieve/pii/S0306261909004279

[14] P. D'Jesús, N. Boukis, B. Kraushaar-Czarnetzki and E. Dinjus. Gasification of corn and clover grass in supercritical water. Fuel [Internet] 2006 [cited 2026 Jan 13];85:1032–8. Available from: https://linkinghub.elsevier.com/retrieve/pii/S0016236105004254

[15] B. Acharya, A. Dutta and P. Basu. Chemical-Looping Gasification of Biomass for Hydrogen-Enriched Gas Production with In-Process Carbon Dioxide Capture. Energy Fuels [Internet] 2009 [cited 2026 Jan 13];23:5077–83. Available from: https://pubs.acs.org/doi/10.1021/ef9003889

[16] Z. Chen, W. Wei, X. Chen, Y. Liu, Y. Shen and B.J. Ni. Upcycling of plastic wastes for hydrogen production: Advances and perspectives. Renew. Sustain.

Energy Rev. [Internet] 2024 [cited 2026 Jan 13];195:114333. Available from: https://linkinghub.elsevier.com/retrieve/pii/S136403212400056X

[17] A. Erkiaga, G. Lopez, M. Amutio, J. Bilbao and M. Olazar. Syngas from steam gasification of polyethylene in a conical spouted bed reactor. Fuel [Internet] 2013 [cited 2026 Jan 13];109:461–9. Available from: https://linkinghub.elsevier.com/retrieve/pii/S001623611300210X

[18] Y. Chai, N. Gao, M. Wang and C. Wu. H2 production from co-pyrolysis/gasification of waste plastics and biomass under novel catalyst Ni-CaO-C. Chem. Eng. J. [Internet] 2020 [cited 2026 Jan 13];382:122947. Available from: https://linkinghub.elsevier.com/retrieve/pii/S1385894719323575

[19] W. Xuan, H. Wang, S. Yan and D. Xia. Exploration on the steam gasification mechanism of waste PE plastics based on ReaxFF-MD and DFT methods. Fuel [Internet] 2022 [cited 2026 Jan 13];315:123121. Available from: https://linkinghub.elsevier.com/retrieve/pii/S001623612102980X

[20] M. He, B. Xiao, Z. Hu, S. Liu, X. Guo and S. Luo. Syngas production from catalytic gasification of waste polyethylene: Influence of temperature on gas yield and composition. Int. J. Hydrog. Energy [Internet] 2009 [cited 2026 Jan 13];34:1342–8. Available from: https://linkinghub.elsevier.com/retrieve/pii/S0360319908016844

[21] V.G. Nguyen, T.X. Nguyen-Thi, Phong P.Q. Nguyen, V.D. Tran, Ü. Ağbulut, L.H. Nguyen, *et al.* Recent advances in hydrogen production from biomass waste with a focus on pyrolysis and gasification. Int. J. Hydrog. Energy [Internet] 2024 [cited 2026 Jan 13];54:127–60. Available from: https://linkinghub.elsevier.com/retrieve/pii/S0360319923023017

[22] A.K. Vuppaladadiyam, S.S.V. Vuppaladadiyam, A. Awasthi, A. Sahoo, S. Rehman, K.K. Pant, *et al.* Biomass pyrolysis: A review on recent advancements and green hydrogen production. Bioresour. Technol. [Internet] 2022 [cited 2026 Jan 13];364:128087. Available from: https://linkinghub.elsevier.com/retrieve/pii/S0960852422014201

[23] G. Lopez, L. Santamaria, A. Lemonidou, S. Zhang, C. Wu, A.T. Sipra, *et al.* Hydrogen generation from biomass by pyrolysis. Nat. Rev. Methods Primer [Internet] 2022 [cited 2026 Jan 13];2:20. Available from: https://www.nature.com/articles/s43586-022-00097-8

[24] Z. Wang, Y. Pan, T. Dong, X. Zhu, T. Kan, L. Yuan, *et al.* Production of hydrogen from catalytic steam reforming of bio-oil using C12A7-O−-based catalysts. Appl. Catal. Gen. [Internet] 2007 [cited 2026 Jan 13];320:24–34. Available from: https://linkinghub.elsevier.com/retrieve/pii/S0926860X06008702

[25] W. Wang. Enhanced hydrogen production using a tandem biomass pyrolysis and plasma reforming process. Fuel Process. Technol. 2022.

[26] G. Duman and J. Yanik. Two-step steam pyrolysis of biomass for hydrogen production. Int. J. Hydrog. Energy [Internet] 2017 [cited 2026 Jan

13];42:17000–8. Available from: https://linkinghub.elsevier.com/retrieve/pii/S0360319917322115

[27] S. Liu, J. Zhu, M. Chen, W. Xin, Z. Yang and L. Kong. Hydrogen production via catalytic pyrolysis of biomass in a two-stage fixed bed reactor system. Int. J. Hydrog. Energy [Internet] 2014 [cited 2026 Jan 13];39:13128–35. Available from: https://linkinghub.elsevier.com/retrieve/pii/S0360319914019016

[28] C.S. Damian, Y. Devarajan, T. Raja and R. Jayabal. A comprehensive review of biomass pyrolysis for hydrogen production in India. Process Saf. Environ. Prot. [Internet] 2024 [cited 2026 Jan 13];190:646–62. Available from: https://linkinghub.elsevier.com/retrieve/pii/S0957582024008619

[29] A.S. Al-Fatesh, N.Y.A. AL-Garadi, A.I. Osman, F.S. Al-Mubaddel, Ibrahim AA, Khan WU, *et al.* From plastic waste pyrolysis to Fuel: Impact of process parameters and material selection on hydrogen production. Fuel [Internet] 2023 [cited 2026 Jan 13];344:128107. Available from: https://linkinghub.elsevier.com/retrieve/pii/S0016236123007202

[30] S. Czernik and R.J. French. Production of Hydrogen from Plastics by Pyrolysis and Catalytic Steam Reform. Energy Fuels [Internet] 2006 [cited 2026 Jan 14];20:754–8. Available from: https://pubs.acs.org/doi/10.1021/ef050354h

[31] D. Yao, H. Yang, H. Chen and P.T. Williams. Co-precipitation, impregnation and so-gel preparation of Ni catalysts for pyrolysis-catalytic steam reforming of waste plastics. Appl. Catal. B Environ. [Internet] 2018 [cited 2026 Jan 13];239:565–77. Available from: https://linkinghub.elsevier.com/retrieve/pii/S0926337318307173

[32] Y. Li, M.A. Nahil and P.T. Williams. Hydrogen/Syngas Production from Different Types of Waste Plastics Using a Sacrificial Tire Char Catalyst via Pyrolysis–Catalytic Steam Reforming. Energy Fuels [Internet] 2023 [cited 2026 Jan 13];37:6661–73. Available from: https://pubs.acs.org/doi/10.1021/acs.energyfuels.3c00499

[33] P.H.M. Putra, S. Rozali, M.F.A. Patah and A. Idris. A review of microwave pyrolysis as a sustainable plastic waste management technique. J. Environ. Manage. [Internet] 2022 [cited 2026 Jan 13];303:114240. Available from: https://linkinghub.elsevier.com/retrieve/pii/S0301479721023021

[34] H. Wang, B. Zhang, P. Luo, K. Huang and Y. Zhou. Simultaneous Achievement of High-Yield Hydrogen and High-Performance Microwave Absorption Materials from Microwave Catalytic Deconstruction of Plastic Waste. Processes [Internet] 2022 [cited 2026 Jan 13];10:782. Available from: https://www.mdpi.com/2227-9717/10/4/782

[35] J. Wang, Y. Pan, J. Song and Q. Huang. A high-quality hydrogen production strategy from waste plastics through microwave-assisted reactions with heterogeneous bimetallic iron/nickel/cerium catalysts. J. Anal. Appl. Pyrolysis [Internet] 2022 [cited 2026 Jan 13];166:105612. Available from: https://linkinghub.elsevier.com/retrieve/pii/S0165237022001826

[36] H.A. Baloch, T. Yang, R. Li, S. Nizamuddin, X. Kai and A.W. Bhutto. Parametric study of co-gasification of ternary blends of rice straw, polyethylene and polyvinylchloride. Clean Technol. Environ. Policy [Internet] 2016 [cited 2026 Jan 13];18:1031–42. Available from: http://link.springer.com/10.1007/s10098-016-1092-4

[37] K.G. Burra and A.K. Gupta. Synergistic effects in steam gasification of combined biomass and plastic waste mixtures. Appl. Energy [Internet] 2018 [cited 2026 Jan 13];211:230–6. Available from: https://linkinghub.elsevier.com/retrieve/pii/S0306261917315702

[38] G. Lopez, A. Erkiaga, M. Amutio, J. Bilbao and M. Olazar. Effect of polyethylene co-feeding in the steam gasification of biomass in a conical spouted bed reactor. Fuel [Internet] 2015 [cited 2026 Jan 13];153:393–401. Available from: https://linkinghub.elsevier.com/retrieve/pii/S0016236115002872

[39] F. Pinto, C. Franco, R.N. André, M. Miranda, I. Gulyurtlu and I. Cabrita. Co-gasification study of biomass mixed with plastic wastes. Fuel [Internet] 2002 [cited 2026 Jan 14];81:291–7. Available from: https://linkinghub.elsevier.com/retrieve/pii/S0016236101001648

[40] J. Alvarez, S. Kumagai, C. Wu, T. Yoshioka, J. Bilbao, M. Olazar, *et al.* Hydrogen production from biomass and plastic mixtures by pyrolysis-gasification. Int. J. Hydrog. Energy [Internet] 2014 [cited 2026 Jan 13]; 39:10883–91. Available from: https://linkinghub.elsevier.com/retrieve/pii/S0360319914012798

[41] X. Liu, K.G. Burra, Z. Wang, J. Li, D. Che, A.K. Gupta. On deconvolution for understanding synergistic effects in co-pyrolysis of pinewood and polypropylene. Appl. Energy [Internet] 2020 [cited 2026 Jan 13];279: 115811. Available from: https://linkinghub.elsevier.com/retrieve/pii/S0306261920312915

[42] F. Pinto, R. André, M. Miranda, D. Neves, F. Varela and J. Santos. Effect of gasification agent on co-gasification of rice production wastes mixtures. Fuel [Internet] 2016 [cited 2026 Jan 13];180:407–16. Available from: https://linkinghub.elsevier.com/retrieve/pii/S0016236116302137

[43] E. Parparita, MdA. Uddin, T. Watanabe, Y. Kato, J. Yanik and C. Vasile. Gas production by steam gasification of polypropylene/biomass waste composites in a dual-bed reactor. J. Mater. Cycles Waste Manag. [Internet] 2015 [cited 2026 Jan 13];17:756–68. Available from: http://link.springer.com/10.1007/s10163-014-0308-0

[44] R. Alipour Moghadam Esfahani, L. Osmieri, S. Specchia, S. Yusup, A. Tavasoli and A. Zamaniyan. H2-rich syngas production through mixed residual biomass and HDPE waste via integrated catalytic gasification and tar cracking plus bio-char upgrading. Chem. Eng. J. [Internet] 2017 [cited 2026 Jan 13];308:578–87. Available from: https://linkinghub.elsevier.com/retrieve/pii/S1385894716312852

[45] J. Chattopadhyay, T.S. Pathak, R. Srivastava andA.C. Singh. Catalytic co-pyrolysis of paper biomass and plastic mixtures (HDPE (high density

polyethylene), PP (polypropylene) and PET (polyethylene terephthalate)) and product analysis. Energy [Internet] 2016 [cited 2026 Jan 13];103:513–21. Available from: https://linkinghub.elsevier.com/retrieve/pii/S0360544216302596

[46] N. Saifuddin, P. Priatharsini, S.B. Hakim. Microwave-Assisted Co-Pyrolysis of Bamboo Biomass with Plastic Waste for Hydrogen-Rich Syngas Production. Am. J. Appl. Sci. [Internet] 2016 [cited 2026 Jan 13];13:511–21. Available from: http://thescipub.com/abstract/10.3844/ajassp.2016.511.521

[47] K. Akubo, M.A. Nahil and P.T. Williams. Co-pyrolysis–catalytic steam reforming of cellulose/lignin with polyethylene/polystyrene for the production of hydrogen. Waste Dispos. Sustain. Energy [Internet] 2020 [cited 2026 Jan 13];2:177–91. Available from: https://link.springer.com/10.1007/s42768-020-00047-8

[48] R. Zou. Catalytic co-pyrolysis of solid wastes (low-density polyethylene and lignocellulosic biomass) over microwave assisted biochar for bio-oil upgrading and hydrogen production. J. Clean. Prod. 2022;

[49] A. Arregi, M. Amutio, G. Lopez, M. Artetxe, J. Alvarez, J. Bilbao, *et al.* Hydrogen-rich gas production by continuous pyrolysis and in-line catalytic reforming of pine wood waste and HDPE mixtures. Energy Convers. Manag. [Internet] 2017 [cited 2026 Jan 14];136:192–201. Available from: https://linkinghub.elsevier.com/retrieve/pii/S0196890417300080

[50] D.B. Levin and R. Chahine. Challenges for renewable hydrogen production from biomass. Int. J. Hydrog. Energy [Internet] 2010 [cited 2026 Jan 13];35:4962–9. Available from: https://linkinghub.elsevier.com/retrieve/pii/S0360319909013561

[51] C. Rodriguez Correa and A. Kruse. Supercritical water gasification of biomass for hydrogen production – Review. J. Supercrit. Fluids [Internet] 2018 [cited 2026 Jan 13];133:573–90. Available from: https://linkinghub.elsevier.com/retrieve/pii/S0896844617303510

[52] K. Erdem, D. Gündüz Han and A. Midilli. A parametric study on hydrogen production by fluidized bed co-gasification of biomass and waste plastics. Int. J. Hydrog. Energy [Internet] 2024 [cited 2026 Jan 13];52:1434–44. Available from: https://linkinghub.elsevier.com/retrieve/pii/S0360319923051819

[53] V. Sridevi, D.V. Surya, B.R. Reddy, M. Shah, R. Gautam, T.H. Kumar, *et al.* Challenges and opportunities in the production of sustainable hydrogen from lignocellulosic biomass using microwave-assisted pyrolysis: A review. Int. J. Hydrog. Energy [Internet] 2024 [cited 2026 Jan 13];52:507–31. Available from: https://linkinghub.elsevier.com/retrieve/pii/S0360319923030902

[54] S.S. Alam and A.H. Khan. Microwave-assisted pyrolysis for waste plastic recycling: A review on critical parameters, benefits, challenges, and scalability perspectives. Int. J. Environ. Sci. Technol. [Internet] 2024 [cited 2026 Jan 13];21:5311–30. Available from: https://link.springer.com/10.1007/s13762-023-05352-3

CHAPTER 12

Importance of Light Weight and High Strength Materials in Modern Society

K. Talukdar
Department of Mechanical Engineering,
Dhemaji Engineering College, Assam, India
E-mail: kamaljyoti.talukdar@gmail.com

ABSTRACT: In modern world population is increasing at fast pace. To fulfill the demand of modern society mass production is done in various industries. Previously traditional materials like pure metals, inorganic-organic compounds, alloys fulfilled the needs. But now due to various demands and depleting resources optimized functional materials have necessitated innovating materials which fulfills two purposes i.e. light weight with high strength at a time. That means by single manufacturing a product it will have dual properties of low/light weight and high strength per unit material. Due to hard and dedicated work of researchers, scientists, and industrialists many materials with these dual properties have been developed. Some of them are alloys of aluminum, titanium, magnesium, and high strength steel which come under metal lists. Similarly some of them are carbon fiber, glass fiber-reinforced polymer composites, carbon nanotubes, graphene, ***metallic microlattices*** *which come under composites. These materials have wide applicability in aerospace industry, automotive industry, construction industry, making sports goods, medical field, and electronics industry. The advantages include enhanced performance due to fuel economy, greater payload capacity, improved speed and operation ability; low cost of manufacturing due to less requirement of material, simple manufacturing process; sustainability due to low energy requirement for manufacturing, reduction of material waste, conservation of resource; flexibility of design due to greater innovative and optimized design options, ability to create complex shapes and structures; safety due to minimization of injury risk and improved crashworthiness. The disadvantages include high cost, high brittleness, difficulty in joining and repairing, poor in durability and structural stability, difficulties in manufacture and availability, poor in recyclability. Although light weight and high strength materials have some drawbacks, with technological advancement it can be reduced by researchers with passage of time.*

Keywords: Composites, Dual Properties, High Strength, Light Weight, Metal.

I. Introduction

Light weight and materials with high strength have gained immense popularity. More such materials are in developing stage.

A. What are light weight and high strength materials?

Materials which have low density and high toughness are called light weight and materials with high strength. The division into two categories are namely metals' list and composites. Materials which are composites are obtained by combining two or more materials having diverse properties. Figure 1 shows various classifications of light weight and materials with high strength.

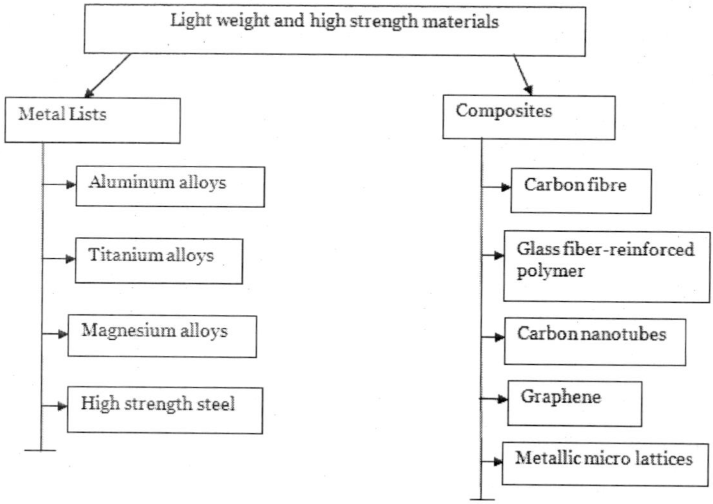

Figure 1: Classification of Light Weight and Matter with High Strength [1]

Table 1 gives an idea of variation of densities and toughness for various materials under metal lists and composites.

Table 1: Values of Density and Hardness

		Density(kg/m³)	*Toughness Value (MPa)*
Metal Lists	Aluminum alloys	2640–2810 [2]	200–600 [3]
	Titanium alloys	4430 [4]	950 [4]
	Magnesium alloys	1800 [5]	135–440 [6]
	High strength steel	7800 [7]	270–700 [8]

		Density(kg/m³)	*Toughness Value (MPa)*
Composites	Carbon fibre	1750 [9]	3500–7000 [10]
	Glass fibre	2540–2780 [11]	3300–4890 [12]
	Carbon nanotube	1300–1400 [13]	11000–63000 [14]
	Graphene	2270 [15]	130000 [15]
	Metallic microlattice	0.9 [16]	100 times of regular polymers [17]

B. Advantages of Light Weight and High Strength Materials

- **Improved performance:** Results lesser fuel consumption in transport sector due to light vehicles; increased carrying load in aerospace, goods carrying vehicles due to light materials; greater speed and manoeuvrability.
- **Reduction of cost of manufacturing:** It is due to lesser usage of materials for same purpose and simplified manufacturing.
- **Greater sustainability:** It is due to lesser energy usage for manufacturing thereby reducing carbon related pollution; less generation of waste and requirement for recycling.
- **Flexibility of design:** These materials give manufacturers greater flexibility in designing innovative and optimized products; helps to create shapes and structures of complex nature.
- **Greater safety:** These materials are easy to handle and operate; greater impact energy absorption capacity ensuring accidents' safety [18].

C. Disadvantages of Light Weight and High Strength Materials

- **Higher cost:** These materials have higher manufacturing cost which impacts large scale productions and its applications.
- **Higher brittleness:** As these materials are more brittle compared to traditional materials, they are not suited for bearing dynamic loads and high impactful forces.
- **Difficulty in joining and repairing:** These materials (composites) are complex to join requiring special methods and instruments. Also, these materials are complex and costly to repair.
- **Lesser durability and stability:** These materials are poor in resisting wind velocities and seismic forces.

- **Complex manufacturing and its availability:** These materials require special process of manufacture with more manufacturing time which may reduce its applicability in certain areas.
- **Problem in recycling:** These materials (composites) are difficult and costly for recycling [19].

II. Applications of Light Weight and High Strength Materials

Light weight with materials having high strength find large applicability in aerospace, automobile, construction industries, and sports, medical, electronics field [1].

A. Application in Aerospace/Aeronautical Industries

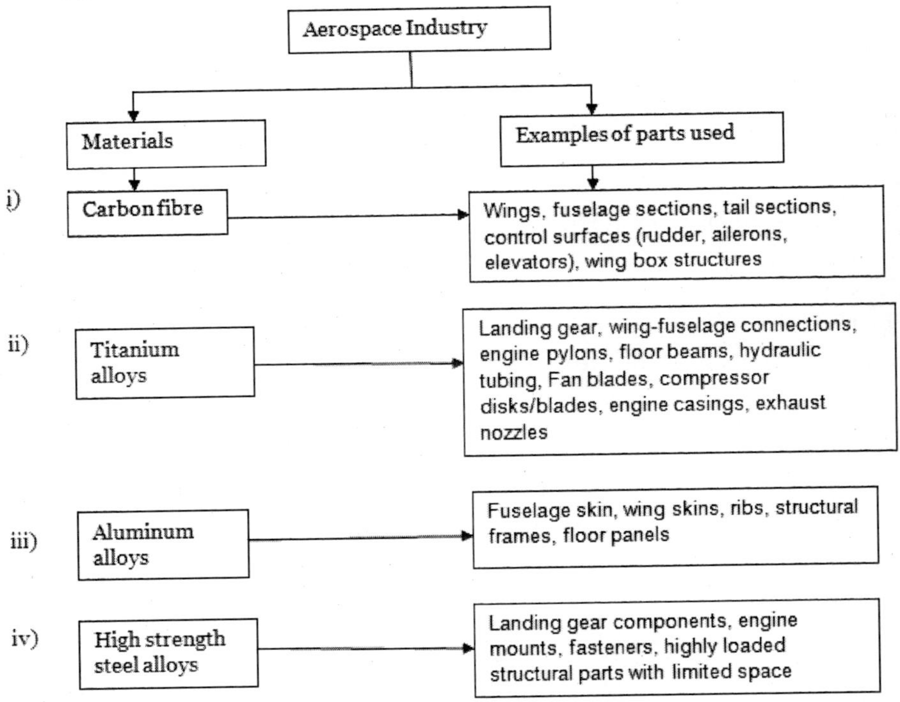

Figure 2: Materials with Associated Parts in Aerospace Industry [20]

Aerospace parts are being made in the advanced manufacturing industry for decreasing consumption due to energy by deviating materials which are light, like composites, polymers, and metals which are lightweight. The usage of metals and composites with lightweight enables satisfying stringent standards in aeronautical applications for weight reduction. The concept of significance of sustainability and study of materials' recyclability with emissions reduction in aeroplanes, and analysis of life cycle were studied. According to the above objectives, Nagaraju, S.B., Priya, H.C. *et al.* presented an overview in condensed form for existing materials which are lightweight and sustainable for applications in aerospace, along with the case studies related to each of those materials. Besides the modern challenges and its future potentiality of lightweight materials were discussed [21].

Williams, J.C. and Starke, E.A examined the development of the role of standpoint in aircraft and aircraft engines that improved materials with processing had done which were customer value driven. It was illustrated that with materials advancement processing technology, clarity had been made possibleby incepting heavier manned air flight. They focused on aircraft and engines of aircraft, with wider focus of materials to create lightweight structures. Some examples were given in the paper having relevance to automotive applications by adjusting the cost [22].

In recent years, significant progress and developments of novel materials for aeronautic applications had been made for reducing costs by increasing engine components structures' running life, fuel economy, capacity to carry more load, and flight's range. Florea, M., Constantin, V.S *et al.* investigated different metallic materials such as alloys of aluminum, titanium, magnesium, steels, superalloys of nickel, and metal matrix composites (MMCs), giving an outline of present advancements and highlights of challenges and future outcomes in aeronautic metals. Various critical factors like withstanding humidity and temperature along with stresses due to tension, compression, bending, cyclic loads, creep, and torsion were taken into consideration along with cost which was critical factor. Trial and error method for material selection could be costly requiring well-planned designing and engineering for ensuring aerodynamic forces' resistance during flight. That approach was used in Boeing 747, Airbus A380, Boeing 787. By use of sandwiched structures research had been enhanced for increasing efficiency of structures and development of materials. Research on sandwiched structures gave excellent stiffness-to-weight ratios and impact energy absorption properties consisting of two thin, rigid face layers bonding to a core material. Besides balsa and foam core materials had been utilized in aviation, with more prevalence on the honeycomb structure which included hexagon, reinforced hexagon, rectangle, flex-core, and square cell, mainly served for supporting loads which are normal oriented in the longitudinal direction and shear loads in the transverse axis [23].

Composite materials are progressing for aerospace structures' construction. During 1960's parts manufactured from materials which are composite like flight controls, spoilers, and fairings, had been developed due to savings in weight compared to parts made from aluminum. Present generation big aircraft are being designed having complete composition of fuselage and winged structures and as its repairement needs detailed structural, materials and tooling knowledge made from composites. The main advantages of composites are high strength, low relative weight, and resistant to corrosion [24].

B. Application in Automobiles

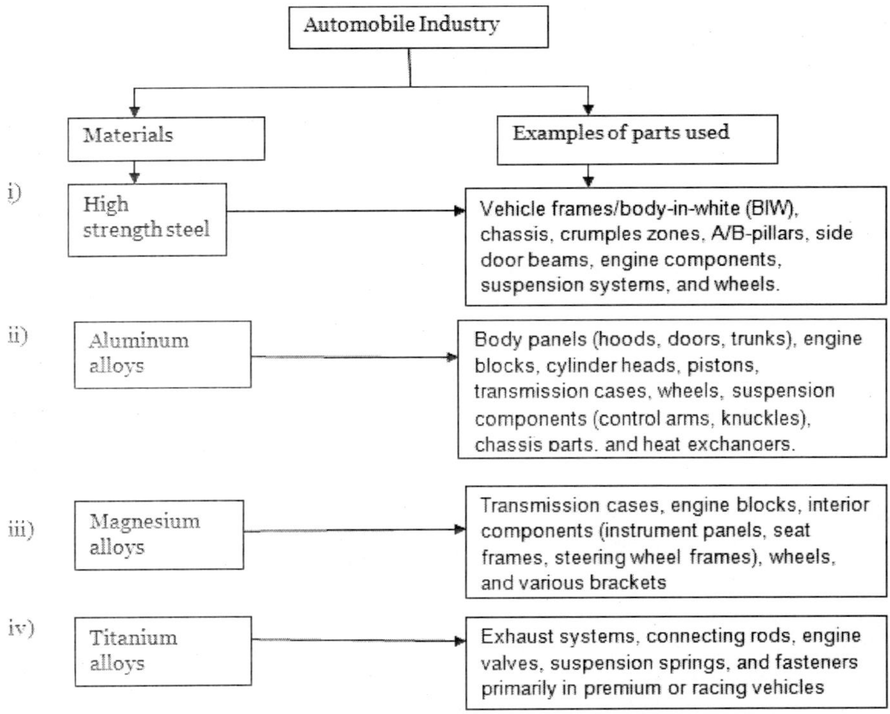

Figure 3: Materials with Associated Parts in Automobile Industry [25]

Due to strict needs for improving economy of fuel and emissions in vehicles, Al and Mg substitution for traditional steel and cast irons has grown. Cole, G. S. and Sherman, A. M addressed some issues related to technical that had to be considered if larger volumes of lightweight materials are to be utilized in the automotive industry. Metals which are lightweight, both cast and wrought forms could be used in the vehicle. Aluminum which are stamped sheet has the capacity for using at a large scale in vehicular structures and closures although in comparison to steel, aluminum has greater difficulty for stamping and spot welding. Al, Mg and castings of metal-matrix composite have the capacity for usage as replacements for numerous castings of ferrous in powertrain and chassis components. But due to numerous tribological features, strength, and ductility, light-metal castings needs procedures related to foundry which are improved with sophisticated design rules before the usage in greater quantities by product engineers. Major challenge for production of component of lightweight materials layed in the function-alibility at an affordable price. The work in this paper concluded with positive cost-benefits for application of light-metals in the automotive industry [26].

The choices of materials for vehicle components were dependent on effectiveness with economy, safeties, and recyclability and lightweight. Earlier for body structure and panels steel, aluminium and magnesium were mainly used. For outer attachments plastics were used. Cars consisting of mainly steel and iron expected reduction of the steel and iron as impending usage in multimaterial constructions. Steel have the probability of substitution by plastics, magnesium and aluminium as alternatives in place of unibody of steel having multimaterial unibodies and space frames of aluminium. Concepts of space frame made of steel and magnesium for applications in volume were in under developed stage. Due to durability, specific strength and stiffness of carbon fibre, traditional materials were replaced [27].

The rapid rise in private cars' number caused to increase energy consumption of China. China is deficit in energy resources. For decreasing the dependency of energy, conservation of energy and reduction of emission from automobiles were given importance. Out of many methods one way was reducing the weight of the car. Upon reduction of car weight, energy consumption gets decreased to achieve the aim of conservation of energy. Han, C. focused on the usage of materials which are lightweight in present automotive materials, and introduced various lightweight materials for weight reduction, consumption of energy, and achieving development in sustainable ways [28].

Two main requirements and improvements of car's body are its lightweight and crashworthiness. For reducing weight of cars without affecting crashworthiness, steel sheet of mild to high strength with reduction of depth was put forwarded. From mechanical analysis depth reducing equation was formed with verification by simulation of fine body parts' element crashed into a rigid wall. Many main parts for impact energy absorption's depth was minimized by bringing forward equation with the simulation results of the car crashing into a rigid wall showing the validity of the presented method [29].

C. Application in Construction Industries

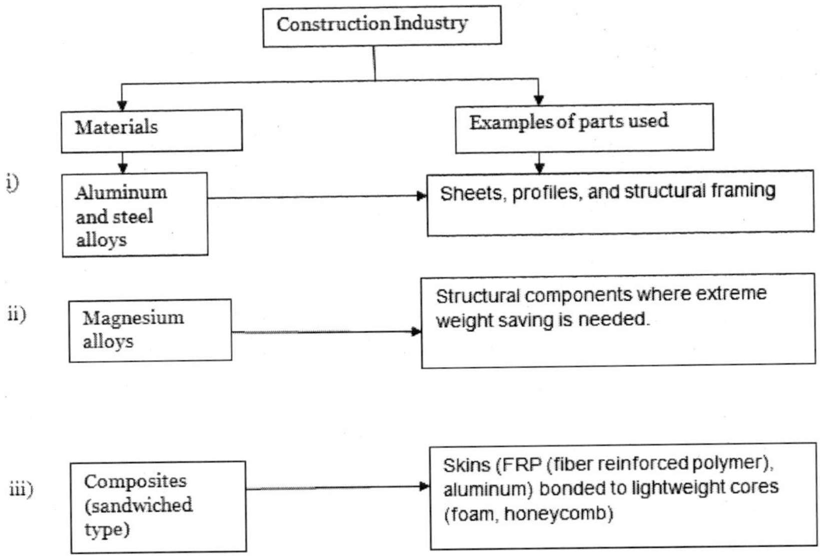

Figure 4: Materials with Associated Parts in Construction Industry [30]

The usage of super-plasticizers with mineral admixtures, namely silica fume which is condensed, had given rise to high strength with concretes of normal weight for applications in structures. The careful usage of such matters gave rise to concrete having compressive strengths greater than 100 MPa. Wilson, H.S. and Malhotra, V.M. developed high strength and lightweight concretes by utilizing lightweight aggregate from Canada. 25 batches having seven concrete mixtured series were made. The density of mixture of silica fume, fly ash and cements varied from 300 to 635 kg/m^3. Entrainment of air and superplasticization were made for all mixtures. For evaluating the mechanical related properties of concrete, freezing and thawing cycles resistance a great number of testing prisms and cylinders were casted, done in accordance to ASTM C 666 Procedure A. Solutions were obtained and inferred that concretes having densities below 2000 kg/m^3 could be structured with or without the usage of admixtures of mineral, without omitting the usage of superplasticizers. For 638 kg/m^3 cementitious density largest compressive strength i.e. 66.5 MPa was found at 365 days. The splitting-tensile strengths of 28 day gave the value of 3.5 MPa. The tests of thawing and freezing were done according to ASTM C 666 Procedure A, freezed in water and thawed in water which gave excellent resistance to frost for concrete having more than 90 durability factors [31].

Presently building industry uses lot of lithospheric resources. For optimum usage a strategy was undergone at the University of Natural Resources and Life Sciences, Vienna and TU Wien for reducing the reinforcement and concrete quantity structured by usage of high-performing materials such as carbon-fibre-reinforced polymers (CFRP) in ultra-high-performance concrete (UHPC) members which permitted a considerable decrease in the weight due to self and allowed the designing of concrete elements of precast type which was very lightweight. The authors used CFRP rods as a tool for bending reinforcement and a flat combination to preform textiles of CFRP for shearing and structural reinforcing. In the 1st half of the paper, introduction of the approach to research having combined state of the art of UHPC and CFRP reinforcement's overview was outlined. The 2nd half contained a concepts related to designing of structural members such as (single floor element and double T-beams), begining with the development of suitable UHPC mixture. Further primary uniaxial tensile tests were described and evaluated for UHPC strips of textile-reinforcement. Finally Kromoser, B., Preinstorfer, P *et al.* did parametric studies related to parameter for various types of reinforcement, its values, rod reinforcement due to additional prestressing and the first prototypes which gave feasibility's estimation, the bearing load with behavior due to deflection of the mentioned components of building [32].

Aluminum, steel or other metallic structures are lighter than building materials which are traditional with 20% weight reduction of the foundation structure. Also these newer high-strength lightweight materials made 50% reduction in energy consumption in buildings which were prefabricated, implying low emissions of CO_2; constructional costs reduction due to material costs reduction, less problems related to waste disposal and improved productivity. Also Liu, J., Huang, M *et al.* reduced the energy requirement for building and maintaining buildings leading to saving money by owners, tenants and contractors [33].

Balendran, R.V., Zhou, F.P *et al.* presented the outcomes of a number of experiments for investigating the fibre inclusion effectiveness for improving mechanical performance of concrete with respect to type and specimen size. The experiments indicated that the small fibre's volume (1%) had small effect on strength which is compressive but showed improved properties on strength which is splitting tensile and flexural along with toughness. The incrementon strength which is splitting tensile and flexural with toughness index for low weight concrete were found comparatively greater than that of concrete of normal aggregate. With increase in specimen size, strength which is splitting and flexural seemed to show decrement, and behavior related to fracture tended to be of higher brittleness [34].

D. Application in Sports Goods

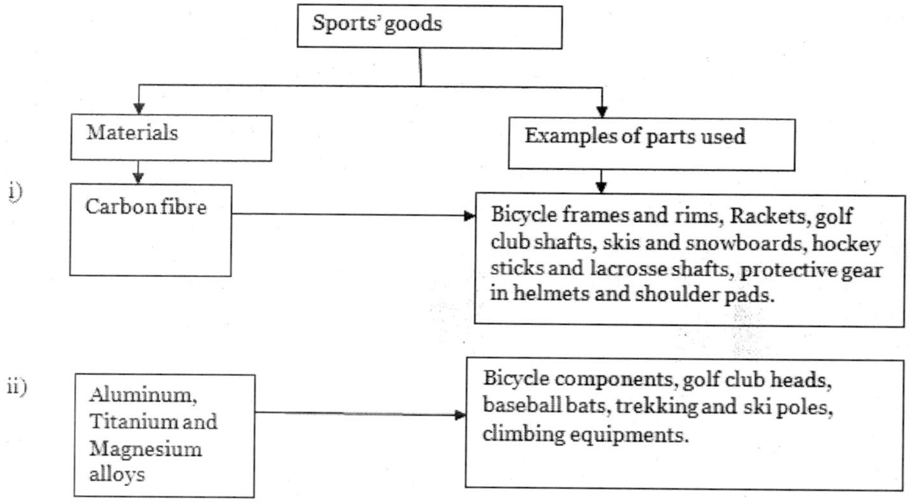

Figure 5: Materials with Associated Parts in Construction Industry [35]

Li Q., Ahmed, I *et al.* provided a simple clarity of polymers, divisions, and its pivotal functions in sports goods and exploring design of low weight, resistance to impact, and modern sports gear. Also procedures for fabrication of numerous composites of polymer matrix had been studied for sports goods. But their study shortly outlined the associated disadvantages with polymers which include cost, durability, longevity, and regular complacency. Finally the study provided probable future outcomes and various role of polymers in sports goods [36].

Bharath, K.N, Joshi, A.G *et al.* focused on the evolution of usage of raw materials i.e. composites for goods and equipment in applications of sports in place of traditional materials like wood, aluminum alloys, etc. having required properties like low weight, more stiffness, strength, with durability which enabled the players to show better results along with safety. Bharath, K.N, Joshi, A.G *et al.* showed that composites of polymer with fiber of synthetic nature being utilized in manufacture of sports equipment. But, their study showed that natural fibers have potentiality for reinforcements as alternate to synthetic fibers. It was anticipated that research and developments prospects were paid more attention to natural composite [37].

Here the applicability of light alloys, aluminum, titanium and magnesium alloys, which were surface engineered in different sports were given introduction, study and discussion [38].

Zhang, J.Z briefed the study of the specific applicability of composite materials made of carbon fiber in making equipments of sports as they are high resistant to temperature, corrosion resistant, low weight and have higher mechanical strength [39].

E. Application in Medical Field

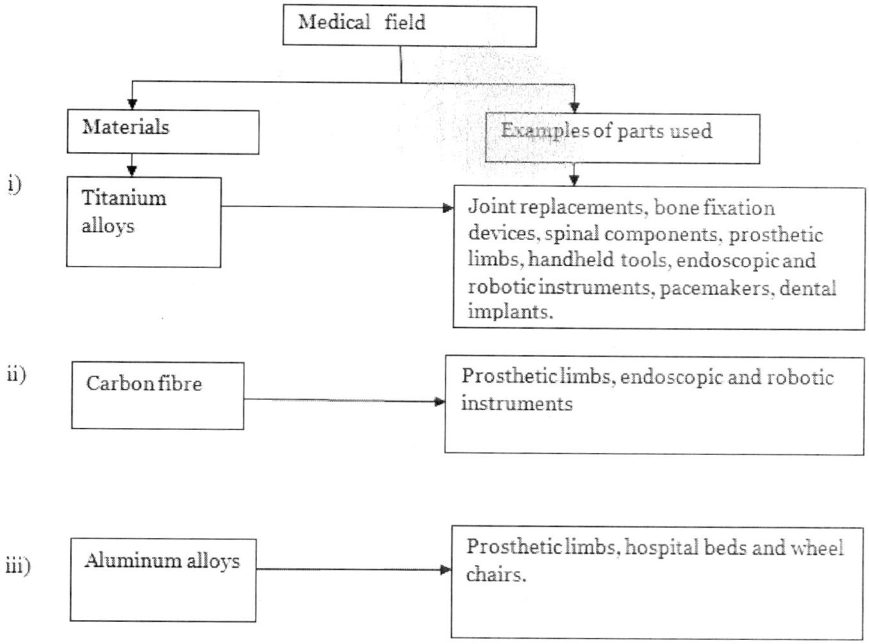

Figure 6: Materials with Associated Parts in Construction Industry [40]

Biomaterials of metal, polymer and ceramic are being in usage for various medical treatments like repairs of tissue along with replacements for past decades. However, composites are gaining recognized. In this article introduction of materials of composite and discussion on recent and potentiality in the biomedical field had been done [41].

All most tissues of human like bones, tendons, skin, ligaments, teeth, etc., are now having composition of composites which are used to make prostheses mimicking biological tissues, matching behavior of mechanical and restoring the functionality of damaged tissue. Salernitano, E. and Migliaresi, C. investigated different composites type already in application for different biomedical applications by outlining along with description of specific advantages and critical issues[42].

Ramakrishna, S., Mayer, J., presented various biomedical applications of polymer-composite materials over last 30 years along with various polymer composite types already being in utilization. Also advantages related to specific usage of biomaterials of polymer-composite in definite applications were stated. Finally the article examined the complex issues and challenges requiring more research and development of materials of polymer composite for improving acceptance in the biomedical industry[43].

Kim, S.C evaluated the results of shielding for fabrics which was shielded and woved by winding of polyethylene (PE) yarn approximately of a 30 μm wire of tungsten with air pressurized dispersion of tungsten coating nanopowder. The experiment's result showed that coated fabric with nanopowder of tungsten improves the performance of shielding for fiber of general tungsten roughly by 15% providing comparatively effective shielded low-dose radiation roughly for X-ray imaging device of 1.2 m. Hence the study inferred that fiber of tungsten may be beneficial for production of low weight shield clothing protecting from radiation scattered in medical institutions[44].

F. Application in Electronics

Shifa, M., Tariq, F *et al.* gave an outline for fabricating and experimenting of multifunctional with hybrid electronic housing composite where radio frequency (RF) was applied in satellite by fabricating a greater sized of hybrid type block of composite type having carbon fiber (CF) and multiwall carbon nanotubes (MWCNTs) with reinforcement in molding technique and epoxy matrix by hand layup and compression. Various tests were performed. Results gave that by adding MWCNT in carbon fibre reinforced panels (CFRP) showed increment in the overall electrical conductivity than without MWCNTs. Also hybrid composite housing undergone test of vibration and when compared with aluminum alloy (AA 6061-T6) housing usually put in structure's of satellite. It was inferred that housing of hybrid composite could be altered with the 50% lighter in weight metallic counterpart for electronics applications in satellite [46].

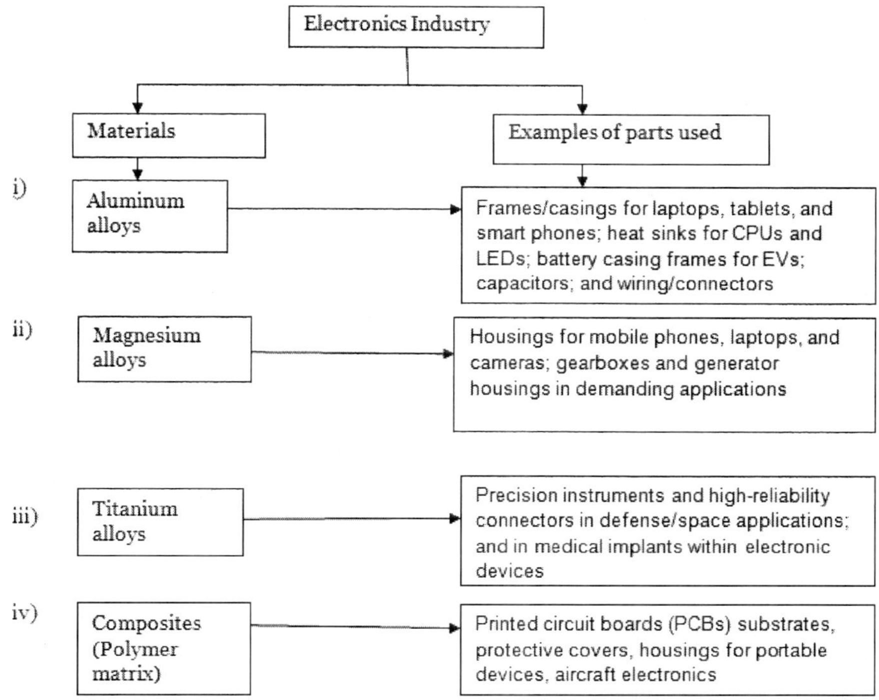

Figure 7: Materials with Associated Parts in Construction Industry [45]

Mounce, S., McPherson, B *et al.* studied the feasibility to develop highly ultra-lightweight SiC with high efficiency, based on DC/DC converter having electrical design, packaging approaches of high-temperature, with testing under high-temperature for number of key components, and packing design at high-temperature. The study would benefit high-temperature environment operated electronics that work outer of spacecraft probes and landers [47].

Ma, M., Qiao, L *et al.* developed a facile method for composites which shields electromagnetic interference (EMI) with strong absorbing and mechanical properties of robust nature by impregnating of suspended graphite nanoplatelets (GNP) and polyacrylate which can expand to foamable particles into the porous reinforced glass mat thermoplastics (GMT), and later pressing and hot forming, upto 2 mm thickness for composites of GMT/PEFP/GNP [48].

Ghosh, S., Ganguly, S *et al.* made a low cost low weight radiation with materials of shielding nature by fabrication by single step, faciled technique without volatile organic solvents of toxic nature by adopting the method of 'dip coating and drying' by preparation of large surface area specialty conductive black (Ketjen-600JD) loaded with composite foam of polyurethane (PU). The stabilized non-ionic surfactant black particles gave improved electrical conductivity with very low percolation threshold concentration and water's durability. It was seen that conducting foam prepared suited for alternative low cost durable EMI shielding [49].

III. CONCLUSION

Finally depending on the above study, it can be said that different applications of low weight and materials of high strength are in infant state. More research should be done to develop such materials and also to reduce the demerits associated with such materials.

REFERENCES

[1] https://hostwebs.site/iGLFK9 (Accessed on 15.8.2025)

[2] The Density of Aluminium and its Alloys. https://www.thyssenkrupp materials.co.uk/density-of-aluminium.html. (Accessed on 7.10. 2025)

[3] Aluminium-Wikipedia. https://en.wikipedia.org/wiki/Aluminium#:~:text=Pure%20aluminium%20is%20quite%20soft,200%20MPa%20to%20600%20MPa. (Accessed on 7.10.2025)

[4] Titanium Ti-6Al-4V - ASM Material Data Sheet – MatWeb. https://asm.matweb.com/search/specificmaterial.asp?bassnum=mtp641.(Accessed on 8.10.2025)

[5] How do magnesium alloys corrode - TFT Pneumatic. https://tft-pneumatic.com/blog/how-magnesium-alloys-corrode/#:~:text=Characteristics%20Of%20Magnesium%20Alloys,density%20is%202400%20kg%2Fm3. (Accessed on 8.10.2025)

[6] Magnesium-alloy–Wikipedia. https://en.wikipedia.org/wiki/Magnesium_alloy#:~:text=Magnesium%20wrought%20alloy%20proof%20stress, AZ61. (Accessed on 8.10.2025)

[7] High-strength low-alloy steel. https://en.wikipedia.org/wiki/High-strength_low-alloy_steel. (Accessed on 8.10.2025)

[8] High Strength Steel vs Advanced High Strength Steel. https://www.aedmetals.com/news/high-strength-steel-vs-advanced-high-strength-steel#:~:text=High%20Strength%20Steel%2C%20a%20type,(270%20to%20700%20MPa). (Accessed on 8.10.2025)

[9] Carbon fiber. https://www.chemeurope.com/en/encyclopedia/Carbon_fiber.html. (Accessed on 8.10.2025)

[10] How strong is carbon fiber. https://carbonfibergear.com/blogs/carbon fiber/how-strong-is-carbonfiber#:~:text=It%20depends%20on%20the%20grade,ksi%20(tensile)%20before%20breaking.(Accessed on 8.10.2025)

[11] https://hostwebs.site/B6CQBN (Accessed on 8.10.2025)

[12] Glass fiber. https://en.wikipedia.org/wiki/Glass_fiber.(Accessed on 8.10. 2025)

[13] Carbon nanotube. https://en.wikipedia.org/wiki/Carbon_nanotube. (Accessed on 8.10.2025)

[14] Mechanical-properties-of-carbon-nanotubes. https://en.wikipedia.org/wiki/ Mechanical_properties_of_carbon_nanotubes.(Accessed on 8.10.2025)

[15] Graphene | Density, Strength, Melting Point - Material Properties. https:// material-properties.org/graphene-density-strength-melting-point/. (Accessed on 8.10.2025)

[16] Metallic microlattice density. https://kg-m3.com/material/metallic-micro-lattice. (Accessed on 8.10.2025)

[17] Metallic-microlattice–Wikipedia. https://en.wikipedia.org/wiki/Metallic_micro lattice#:~:text=The%20material%20is%20described%20as,times%20stronger% 20than%20regular%20polymers. (Accessed on 8.10.2025)

[18] https://hostwebs.site/8IBfNY (Accessed on 16.8.2025)

[19] https://hostwebs.site/q5qJPy (Accessed on 16.8.2025)

[20] https://hostwebs.site/NQR2Ik (Accessed on 16.12.2025)

[21] Nagaraju, S. B., Priya, H. C., Girijappa, Y. G. T. and Puttegowda, M., 9 - Lightweight and sustainable materials for aerospace applications, Lightweight and Sustainable Composite Materials, Preparation, Properties and Applications, 157–178 (2023).

[22] Williams, J.C. and Starke, E.A., Progress in structural materials for aerospace systems, Acta Materialia, 51(19), 5775–5799 (2003).

[23] Florea, M., Constantin, V.S., Bucur, A. C., Faur, R., Predu, D. and Cazacu, A., Materials and Structures Used in Aeronautics: Present and Future Perspectives, INCAS BULLETIN, 16(4), 57–72 (2024).

[24] A. Kalaiyarasan, P. Ramesh, and P. Paramasivam, Study of Advanced Composite Materials in Aerospace Application, International Journal of Scientific Research in Mechanical and Materials Engineering, 2(1), 1–10 (2018).

[25] https://hostwebs.site/NWoaWh (Accessed on 16.12.2025)

[26] G.S. Cole and A.M. Sherman, Light weight materials for automotive applications, Materials Characterization, 35(1), 3–9 (1995).

[27] H. Saidpour, Lightweight High Performance Materials for Car Body Structures, NTI Technology Conference, CEME, Ford Motor Company, USA, 14–19 (2004).

[28] C. Han, Research on the Development and Application of Lightweight Automotive Materials, Journal of Physics: Conference Series-6th International Conference on Materials, Mechanical Engineering and Automation Technology, China, 1676, 1–4 (2020).

[29] Y. Li, Z. Lin., A. Jiang and G. Chen., Use of high strength steel sheet for light weight and crashworthy car body, Materials and Design, 24(3), 177–182 (2003).

[30] https://hostwebs.site/w7qGe0 (Accessed on 16. 12.2025)

[31] H.S. Wilson and V.M. Malhotra, Development of high strength lightweight concrete for structural applications, International Journal of Cement Composites and Lightweight Concrete, 10(2), 79–90 (1988).

[32] B. Kromoser, P. Preinstorfer and J. Kollegge, Building lightweight structures with carbon-fiber-reinforced polymer-reinforced ultra-high-performance concrete: Research approach, construction materials, and conceptual design of three building components, Structural Concrete, 20(2), 730–744 (2018).

[33] J. Liu, M. Huang., S. Li and J. Chen, Application of New Light and High Strength Materials in Prefabricated Buildings, Innovative Computing Vol 1 - Emerging Topics in Artificial Intelligence. IC 2023. Lecture Notes in Electrical Engineering, Springer, Singapore 1044, 951–957 (2023).

[34] R.V. Balendran., F.P. Zhou, A. Nadeem and A.Y.T. Leung, Influence of steel fibres on strength and ductility of normal and lightweight high strength concrete, Building and Environment, 37(12), 1361–1367 (2002).

[35] https://hostwebs.site/A3zdPU (Accessed on 16.12.2025)

[36] Q. Li, I. Ahmed, P.M. Ngoc, T.P. Hoa, T.V. Dieu, M.S. Irshad, H.X. Nang and V.D. Dao, Contemporary advances in polymer applications for sporting goods: fundamentals, properties, and applications, RSC Adv., 14, 37445–37469 (2024).

[37] K.N. Bharath, A.G. Joshi, A. Taj, P.S.S. Gouda and M. Nagamadhu, Chapter 9 – Evolution and recent advancements of composite materials in sports applications, Applications of Composite Materials in Engineering, Woodhead Series in Materials, 219–228 (2025).

[38] J. Chen, 16 – Surface engineered light alloys for sports equipment, Surface Engineering of Light Alloys, Aluminium, Magnesium and Titanium Alloy, Woodhead Publishing Series in Metals and Surface Engineering, 549–567(2010).

[39] J.Z. Zhang, Study on carbon fiber composite materials in sports equipment, Applied Mechanics and Materials, 329, 105–108 (2013).

[40] https://hostwebs.site/KqKEnL (Accessed on 16.12.2025)

[41] M.K. Egbo, A fundamental review on composite materials and some of their applications in biomedical engineering, Journal of King Saud University – Engineering Sciences, 33(8), 557–568 (2021).

[42] E. Salernitano and C. Migliaresi, Composite Materials for Biomedical Applications: A Review, Journal of Applied Biomaterials and Biomechanics, 1(1), 3–18 (2003).

[43] S. Ramakrishna, J. Mayer, E. Wintermantel and W. Kam, K.W. Leong., Biomedical applications of polymer-composite materials: a review, Composites Science and Technology, 61(9), 1189–1224 (2001).

[44] S.C. Kim, Development of a Lightweight Tungsten Shielding Fiber That Can Be Used for Improving the Performance of Medical Radiation Shields, Applied Sciences, 11(14), 6475(2021).

[45] https://hostwebs.site/wYzZPt (Accessed on 16.12.2025)

[46] M. Shifa, F. Tariq, F. Khan, Z.S. Toor and R.A. Baloch, Towards light weight multifunctional hybrid composite housing for satellite electronics, Materials Research Express, IOP Publishing Ltd, 6(12), 125629 (2020).

[47] S. Mounce, B. McPherson, R. Schupbach and A.B. Lostetter, Ultra-lightweight, high efficiency SiC based power electronic converters for extreme environments, IEEE Aerospace Conference (2006).

[48] M. Ma, L. Qiao, H. Yu, S. Chen, Y. Shi, H. He and X. Wang, Lightweight and high-strength GMT/PEFP/GNP composites with absorb-dominated electromagnetic interference shielding property, J. Mater. Sci.: Mater. Electron., 32, 25863–25875 (2021).

[49] S. Ghosh, S. Ganguly, S. Remanan, S. Mondal, S. Jana, P.K. Maji, N. Singha and N.C. Das, Ultra-light weight, water durable and flexible highly electrical conductive polyurethane foam for superior electromagnetic interference shielding materials, Journal of Materials Science: Materials in Electronics, 29(2), 10177–10189 (2018).

CHAPTER 13

Europium (III) Doped Zirconia Nanophosphor: Development of a Potential White Light Source

S.D. Meetei

Department of Physics, Dhanamanjuri University, Imphal-West, Manipur, India

E-mail: sdmdhiren@gmail.com

ABSTRACT: In the present scenario of efficient white light sources for future lighting, europium (III) doped zirconia (ZrO_2: Eu^{3+}) nanophosphors attracted many researchers. It can be affectively, easily and economically synthesized by chemical routes. Hydrothermal method of chemical synthesis is found to be a promising method for synthesis of high purity nano-sized zirconia. Among the various techniques of material characterization, Fourier transform infrared spectroscopy (FTIR) is an indispensable tool for characterization of tetragonal and cubic zirconia. The Comission Internationale de l'éclairage (CIE) color space co-ordinate for undoped sample show within greenish blue color. However, the co-ordinates of the doped samples are found within the region of white color. The greenish blue color emitted from the host ZrO_2 combined with the well-known red color emitted from Eu^{3+} which gives rise to white color. This white light emitting phosphor have low correlated color temperature (CCT) and has potentials for applications in various devices such as in white LEDs.

Keywords: Europium (III) Doped Zirconia; Nanophosphor; White Light Source, Cubic, Low CCT.

I. INTRODUCTION

Lanthanide ion doped nanomaterials continue to remain as an active area of research because of its novel physical and chemical properties which are different from their bulk counterparts. Nanocrystalline zirconium dioxide or zirconia (ZrO_2) doped with Eu^{3+} is one such material and has aroused great interest due to its various applications [1-4]. Zirconia is widely used as refractory, catalyst, oxygen sensor, fuel cell [4, 5], nucleic acid sensor [6], bone cement additives [7] and super hard materials [8]. Because of its high melting point, high thermal and mechanical resistance, high thermal expansion coefficient, low thermal conductivity, high thermochemical resistance, high corrosion resistance, high dielectric constant and photothermal stability, it has extensive applications in photonics and other industries [9-12]. High chemical and photochemical stability with high refractive index and low phonon energy makes it an ideal medium for preparation of highly luminescent materials [9, 11-13]. In luminescence, white light emitting phosphors are of great interest because of its tremendous application in various forms of lighting devices. Many reports of white light emission from materials containing more than one lanthanide ions are available in literatures [14-17]. However, such multiple doped materials have problems of self-absorption thereby reducing their performance. Therefore, single lanthanide ion doped materials are reported to be potential candidates for this purpose [18, 19]. On the other hand, producing white light by doping a single lanthanide ion is seldom reported. Therefore, the development of $ZrO_2:Eu^{3+}$ as a potential white light source may attract significant attention in current efforts to find efficient white light sources.

II. Synthesis

As a luminescent material, properties of zirconia can be tuned with synthesis technique whereby manipulating the size of the particle, number of dopant(s), etc. For this purpose, different synthesis process like chemical, physical methods, etc., can be employed. Among them synthesis by chemical technique is preferred most. Because, it is economical and can easily manipulate the particle size. It is to be noted that optical, electrical, mechanical properties, etc., of a material change as the size of the particle's changes [18, 19]. Moreover, especially for introducing a dopant in a host lattice as in the case of $ZrO_2:Eu^{3+}$, this technique is preferred most. The chemical method for synthesis can also be classified into precipitation, polyol, hydrothermal, etc. according to the type of reactant used, synthesis process employed, etc. Among them precipitation method is considered to be one of the easiest and economical for synthesis of zirconia nanomaterial. The size of the particle can easily be varied in nanometer range, by this method [18, 19]. However, the crystalline phase of zirconia obtained by this synthesis method is confined to monoclinic and tetragonal phases. The most desired cubic phase cannot be obtained by this method. It is found in report that application of zirconia devices solely depends on the cubic phase [19]. Therefore, nanomaterial zirconia is to be synthesized by some other techniques so that the crystalline phase is cubic. One promising technique which can accomplishes the cubic phase is polyol method of synthesis [19]. In this method polyethylene glycol and/or ethylene glycol are/is used. Through this method, cubic zirconia can be obtained without compromising the size. However, the precursors are found to be in gel form and crystallization occurs at high temperature (at 500°C). Therefore, annealing at high temperature is needed to crystallize. On the other hand, formation of crystals as-synthesis is desired. Therefore, there is a need of exploring different methods of zirconia synthesis other than by precipitation and polyol methods. In this regard, hydrothermal method of synthesis is found to be a promising technique.

It is found in the reports that materials synthesized by hydrothermal technique have high purity [20]. With this method of synthesis cubic phase of zirconia is easily obtained as-prepared (without annealing). However, this method required high pressure to crystallize the particle. Fortunately, this can be achieved by using autoclave. Therefore, undoped and europium (iii) doped zirconia sample were synthesized by using hydrothermal technique reported in Ref. [21].

III. CHARACTERIZATIONS

Among the various techniques of nanomaterial characterization, x-ray diffraction (XRD), Fourier transform infra-red (FTIR), transmission electron microscopy (TEM) and its selected area electron diffraction (SAED) are most common and essential [18,19]. Crystalline phase identification of zirconia is done by XRD. Among the three polymorphs of zirconia i.e., monoclinic, tetragonal and cubic phases; identification of crystalline phase of tetragonal and cubic phases solely from the XRD can be misleading. Because the XRD patterns of tetragonal and cubic phases of zirconia are almost similar. However, diffraction rings of a SAED image gives clue for the zirconia material to be tetragonal or cubic [19]. Interestingly, vivid difference can be easily obtained from FTIR spectra. In this point of view FTIR plays an important role in characterizing tetragonal and cubic zirconia. FTIR spectra which shows the difference between tetragonal and cubic zirconia can be found elsewhere [19].

IV. PHOTOLUMINESCENCE

PL emission spectrum of ZrO_2: Eu^{3+} is shown in Figure 1. With addition to the strong emission peak at 470 nm, which is due to the host ZrO_2, Eu^{3+} peaks are also observed at 597 and 613 nm. The peak at 597 nm results from magnetic dipole transition (M), $^5D_0 \rightarrow {}^7F_1$. And the peak at 613 nm results from structurally sensitive electric dipole transition (E), $^5D_0 \rightarrow {}^7F_2$. The relative intensity of them depends strongly on the local symmetry of Eu^{3+} and a lower symmetry of crystal field around Eu^{3+} will result in better chromaticity [22-27].

Figure 1 also clearly shows the asymmetry ratio (E/M) of the sample. It is also found in reports that, monoclinic phase has lower symmetry than tetragonal phase [25-26]. And monoclinic phase domination over tetragonal phase will be further confirmed by XRD analysis.

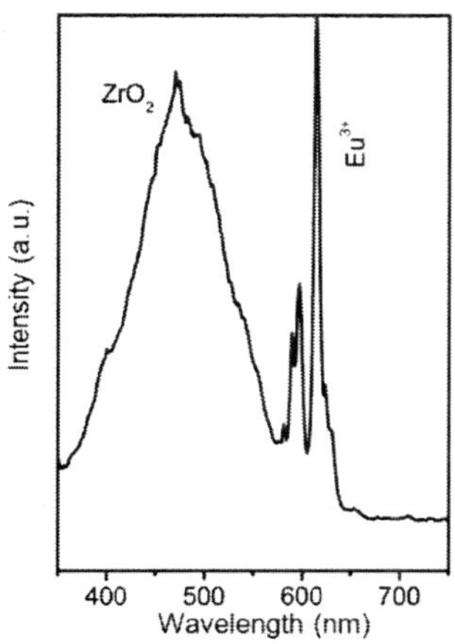

Figure 1: PL Emission Spectrum of ZrO_2:Eu^{3+} Showing Emission from the Host ZrO_2 and Dopant Eu^{3+}

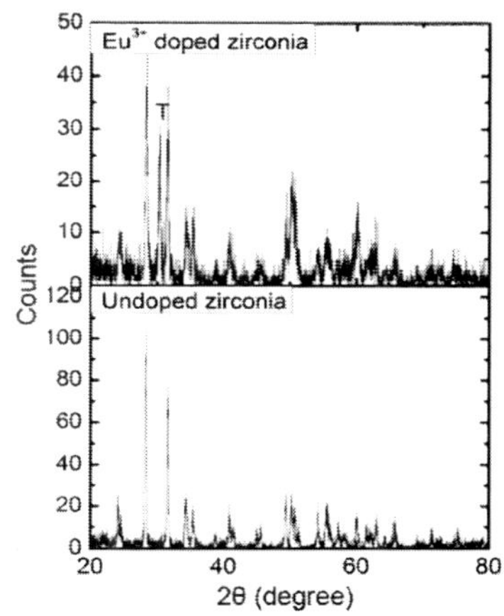

Figure 2: XRD Pattern of Undoped ZrO_2 and Eu^{3+} Doped ZrO_2. On Doping Eu^{3+} in ZrO_2, the Intensity of Tetragonal Peak (T) is Observed

V. Photoluminescence and Crystal Structure Correlation

Undoped zirconia is found in monoclinic phase. However, on introduction of Eu^{3+} in ZrO_2, the intensity of tetragonal peak is observed. This is shown in XRD pattern of ZrO_2 and ZrO_2:Eu^{3+} shown in Figure 2, and marked it 'T'. Appearance of the tetragonal phase can be explained on the basis of oxygen vacancy evolved when Zr^{4+} is replaced by Eu^{3+}. It is to be noted that Eu^{2+} did not formed in the ZrO_2. This is because Eu^{3+} has negative charge relative to ZrO_2 [24]. The appropriate place for Eu^{3+} in the ZrO_2 is Zr^{4+} site and the charge difference is compensated by an oxygen vacancy (O^{2-}) [28, 29]. Since, one oxygen vacancy serves to compensate the difference in charge between Eu^{3+} and Zr^{4+}, one oxygen vacancy is evolved for every two Zr^{4+} replaced by two Eu^{3+} ions.

VI. White Light Emission

Now-a-days, white light emitting phosphors (WLEPs) are of great interest to researchers for it has fantastic uses in various forms of lighting devices such as in white light emitting diodes (WLEDs). With this phosphor, it is anticipated to replace traditional incandescent and fluorescent lamps for general illumination purposes [30-36]. Consequently, present day luminescence studies mainly focused on synthesis of WLEPs for WLEDs. Most of reports synthesized WLEPs through a solid-state reaction technique [30-36]. However, this technique suffers serious drawbacks such as reactants should mix to a homogenous system, high viscosity in reactant system, unsuitable for solvent assisted chemical reactions, among others [37-40]. Moreover, most of the reported phosphors for applications in WLEDs are complex in nature and consequently complex in synthesis [30-35]. As a result, simple aqueous precipitation route for synthesis of WLEPs are preferred over solid-state reaction.

The function of phosphors is to absorb ultraviolet (UV) or blue LED chip and convert it to visible light [30,40]. Thus, for potential use, phosphors should have strong excitation at ultraviolet and should emit the desired light [11]. However, LEDs pumped by near-UV (NUV) influences the purity of white color [36]. Furthermore, lifetime of red phosphor is inadequate under NUV irradiation [32]. And hence, invisible UV (<380 nm) pumped white light with adequate lifetime is required to enhance the luminescence efficiency [36]. Moreover, higher radiative output can be obtained from UV pumped LEDs and production of UV is easier [41]. In view of these facts, most luminescence studies have concentrated on the excitation of phosphors by UV [30, 36, 40-42]. ZrO_2 has strong absorption in the UV region and emission in the visible region and it indicates its potentials for application in white light emission.

The most common and commercialized WLEDs consists of a blue InGaN chip and a yellow cerium doped yttrium aluminum garnet ($YAG:Ce^{3+}$) phosphors [32-34]. However, such phosphors suffer from the lack of a red component [32-34]. And this can be rectified by the used of trichromatic (BGR) phosphors. Unfortunately, the red emitting component of the tricolor phosphor is chemically unstable and shows luminescence saturation [31]. Moreover, emission intensity of the red emitting phosphor is low relative to blue and green counterparts [40, 43]. Therefore, stable red-emitting phosphors with high luminescence efficiency and intensity are necessary for both the methods [31, 32, 44]. It is to be noted that, Eu^{2+} doped phosphors suffer from a difficult preparation process, poor absorption in the blue light and short lifetime [30, 34, 45]. In this point of views, current research for red phosphors has been focused on Eu^{3+} doped materials [34, 42, 45, 46]. Accordingly, it is an attractive and challenging research task to develop Eu^{3+} doped zirconia, a novel and stable red phosphor with comparatively modest synthesis technique that can be pumped effectively by UV, having strong emission comparable to blue and green counterpart with adequate lifetime.

From the PL emission data of this phosphor, the *Commission Internationale de l'éclairage* (CIE) color space co-ordinates are calculated. The CIE values of the doped sample is found to be (0.33, 0.32). This is very close to that of ideal white light (0.33, 0.33). The color space co-ordinates for undoped ZrO_2, recorded under UV excitation is (0.12, 0.23) which corresponds to greenish blue color of the chromaticity diagram. The color space co-ordinates corresponding to undoped and Eu^{3+} doped ZrO_2 are shown in Figure 3.

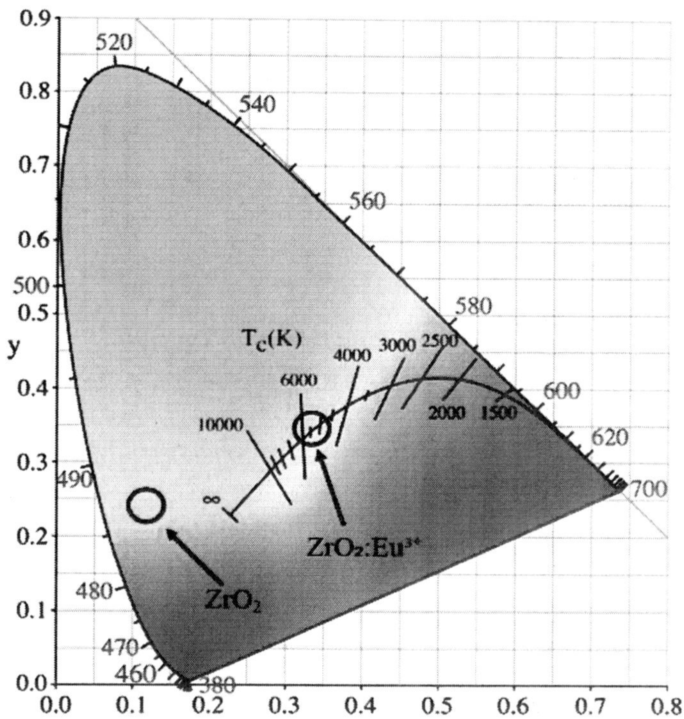

Figure 3: CIE Color Space Co-ordinates Showing Colors of Undoped ZrO_2 and Eu^{3+} Doped ZrO_2. The Figure also Shows Chromaticity of Black-body Light Sources of Various Temperatures, and Lines of Constant CCT

The CIE co-ordinates (*x, y*) of the doped samples shows that the greenish blue color emitted from the host zirconia combined with the well-known red color emitted from the dopant Eu^{3+} which results in white light emission from these phosphors. Therefore, it is reported that the problems which results from the mixing of three different phosphors for three different colors (BGR) can be resolved by using this single dopant phosphors. The white light emission recorded under 245 nm excitation wavelength gives better (*x, y*) values than that of 288 nm excitation wavelength. This may be due to comparable intensity of the emission peaks due to the ZrO_2 and Eu^{3+} and similar report is found [42]. This WLEP do not suffer from strong change in chromaticity. This is because intra-rare-earth atomic transitions occurring in the phosphor do not depend on temperature [39]. Moreover, the white light emitted from this Eu^{3+} doped ZrO_2 is found to have low correlated color temperature (~5500 K) as it is shown in Figure 3. For technical applications, phosphor which generates white light with low correlated color temperatures (CCT) is desirable [47-49].

VII. Conclusion

Chemical synthesis can be employed effectively for the synthesis of ZrO_2 and Eu^{3+} doped ZrO_2 nanophosphors. Among the chemical methods of nanomaterial synthesis, hydrothermal technique is considered to be one of the best methods for synthesizing luminescent zirconia. Characterization by FTIR is indispensable, especially while dealing with tetragonal and/or cubic phase of zirconia. The cubic phase is considered to be the most desired phase of zirconia. The phases of zirconia relate to its photoluminescence properties. ZrO_2: Eu^{3+} is found to be a potential candidate for development of white light. The white light ZrO_2: Eu^{3+} have low CCT. Therefore, study on this material will give a significant development of future solid-state lighting such as in WLEDs.

References

[1] H.C. Madhusudhana and C. Manjunath, Spectroscopic Studies of Red Emitting ZrO_2:Eu^{3+} Nanophosphors using Judd–Ofelt Theory for Lighting Applications, *J. Fluoresc,* (doi: 10.1007/s10895-025-04316-9; online ahead of print) (2025).

[2] I. Kumar, A. Kumar and A.K. Gathania., Activation of zirconia nanophosphors with Eu^{3+} to demonstrate multifunctional optical applications, *Mater. Chem. Phys.,* 292, 126846 (2022).

[3] A.A. Nashivochnikov, A.I. Kostyukov, M.I. Rakhmanova, S.V. Cherepanova and V.N. Snytnikov, Photoluminescence and structure evolution of laser synthesized $ZrO_2:Eu^{3+}$ nanopowders depending on the dopant concentration, *Ceram. Int.*, 49(3), 5049–5057 (2023).

[4] L. Kumari, W. Li and D. Wang, Monoclinic zirconium oxide nanostructures synthesized by a hydrothermal route, *Nanotechnology*, 19, 195602 (2008).

[5] H. Cao, X. Qiu, B. Luo, Y. Liang, Y. Zhang, R. Tan, M. Zhao and Q.M. Zhu, Synthesis and room-temperature ultraviolet photoluminescence properties of zirconia nanowires, *Adv. Funct. Mater.*, 14(3), 243–246 (2004).

[6] E.E. Bolukbas, Ü.H. Kaynar, I. Gozde Kaptanoglu, S. Yusan and D. Özkan-Ariksoysal, Development of undoped or doped zinc and zirconium-based metal oxides for voltammetric sensing of DNA, *Mater. Sci. Eng. B*, 319, 118377 (2025).

[7] Wikipedia, Zirconium dioxide, [http://en.wikipedia.org/wiki/Zirconium_dioxide] (Last accessed, 5 Sept., 2025).

[8] M. Das, G. Sumana, R. Nagarajan and B.D. Malhotra, Zirconia based nucleic acid sensor for Mycobacterium tuberculosis detection, *Appl. Phys. Lett.*, 96, 133703 (2010).

[9] R. Gillani, B. Ercan, A. Qiao and T.J. Webster, Nanofunctionalized zirconia and barium sulfate particles as bone cement additives, *Int. J. Nanomedicine*, 5, 1–11 (2010).

[10] Y. Al-Khatatbeh, K.K.M. Lee and B. Kiefer, Phase relations and hardness trends of ZrO_2 phases at high pressure, *Phys. Rev. B*, 81, 214102 (2010).

[11] E. De la Rosa, L.A. Díaz-Torres, P. Salas and R.A. Rodríguez, Visible-light emission under UV and IR excitation of rare earth-doped ZrO_2 nanophosphor, *Opt. Mater.*, 27, 1320–1325 (2005).

[12] H.D.E. Harrison, N.T. McLamed and E.C. Subbarao, A new family of self-activated phosphors, *J. Electrochem. Soc.*, 110, 23–28 (1963).

[13] P. Ghosh and A. Patra., Role of surface coating in ZrO_2/Eu^{3+} nanocrystals, *Langmuir*, 22(14), 6321–6327 (2006).

[14] M.N. Luwang, R.S. Ningthoujam, S.K. Srivastava. and R.K. Vatsa, Preparation of white-light emitting $YVO_4:Ln^{3+}$ and silica-coated $YVO_4:Ln^{3+}$ nano-particles—energy transfer and site symmetry studies, *J. Mater. Chem.*, 21, 5326-5337 (2011).

[15] S. Sivakumar, F.C.J.M. van Veggel and M. Raudsepp, Bright white light through up-conversion of a single NIR source from sol–gel-derived thin film made with Ln^{3+}-doped LaF_3 nanoparticles, J. Am. Chem. Soc., 127(36), 12464–12465 (2005).

[16] S.H. Lee, J.H. Park, S.M. Son, J.S. Kim and H.L. Park, White-light-emitting phosphor: $CaMgSi_2O_6:Eu^{2+}$, Mn^{2+} and its related properties with blending, *Appl. Phys. Lett.*, 89, 221916 (2006).

[17] L. Chen, K. Chen, S. Hu and R. Liu, Combinatorial chemistry approach to searching phosphors for white light-emitting diodes in (Gd-Y-Bi-Eu)VO_4 quaternary system, *J. Mater. Chem.*, 21, 3677–3685 (2011).

[18] S.D. Meetei, S.D. Singh, N.S. Singh, V. Sudarsan, R.S. Ningthoujam, M. Tyagi, S.C. Gadkari, R. Tewari and R.K. Vatsa, Crystal structure and photoluminescence correlations in white emitting nanocrystalline ZrO_2:Eu^{3+} phosphor: Effect of doping and annealing, *J. Lumin.*, 132, 537–544 (2012).

[19] S.D. Meetei, S.D. Singh and V. Sudarsan, Polyol synthesis and characterizations of cubic ZrO_2:Eu^{3+} nanocrystals, J. Alloys Compd., 514, 174–178 (2012).

[20] R. Piţicescu, C. Monty and D. Millers, Hydrothermal synthesis of nanostructured zirconia materials: Present state and future prospects, *Sens. Actuators B*, 109, 102–106 (2005).

[21] S.D. Meetei and S.D. Singh, Hydrothermal synthesis and white light emission of cubic ZrO_2:Eu^{3+} Nanocrystals, J. Alloys Compd., 587, 143 (2014).

[22] S. Sapra, S. Mayilo, T.A. Klar, A.L. Rogach and J. Feldmann, Bright white-light emission from semiconductor nanocrystals: by chance and by design, *Adv. Mater.*, 19, 569 (2007).

[23] T. Trindade, P. O'Brien and N.L. Pickett, Nanocrystalline Semiconductors: Synthesis, Properties, and Perspectives, *Chem. Mater.*, 13(11), 3843–3858 (2001).

[24] T. Ninjbadgar, G. Garnweitner, A. Borger, L.M. Goldenberg, O.V. Sakhno and J. Stumpe, Synthesis of luminescent ZrO_2:Eu^{3+} nanoparticles and their holographic sub-micrometer patterning in polymer composites, *Adv. Funct. Mater.*, 19(11), 1819–1825 (2009).

[25] K. Smits, L. Grigorjeva, D. Millers, A. Sarakovskis, A. Opalińska, J.D. Fidelus and W. Lojkowski, Europium-doped zirconia luminescence, *Opt. Mater.*, 32, 827–831 (2010).

[26] L. Chen, Y. Liu and Y. Li, Preparation and characterization of ZrO_2:Eu^{3+} phosphors, *J. Alloys Compd.*, 381(1–2), 266–271 (2004).

[27] Wikipedia, Light-emitting diode — Phosphor-based LEDs, [http://en.wikipedia.org/wiki/Light-emitting_diode#Phosphor-based_LEDs] (Last accessed, 5 Sept., 2025).

[28] D. Simeone, G. Baldinozzi, D. Gosset, M. Dutheil, A. Bulou and T. Hansen, Monoclinic to tetragonal semireconstructive phase transition of zirconia, *Phys. Rev. B*, 67, 064111 (2003).

[29] D. Millers, K. Smits, L. Grigorjeva and J.D. Fidelus, Luminescence of Europium-doped ZrO_2 nanopowder, E-MRS Fall Meeting (conference poster) (2006).

[30] H. He, R. Fu, X. Song, D. Wang and J. Chen, White light-emitting $Mg_{0.1}Sr_{1.9}SiO_4$:Eu^{2+} phosphors, J. Lumin., 128(3), 489–493 (2008).

[31] P. Dai, X. Zhang, X. Li, G. Wang, C. Zhao and Y. Liu, Red-emitting $LiEuMo_{2-x}Si_xO_8$ phosphors for white light-emitting diodes, J. Lumin., 131(4), 653–656 (2011).

[32] Y. Hong, Gao (Guimei), K. Li, Liu (Guanghuan), Gan (Shucai) and G. Hong, Synthesis and luminescence properties of a novel red-emitting phosphor $SrCaSiO_4:Eu^{3+}$ for ultraviolet white LEDs, J. Rare Earths, 29, 431–435 (2011).

[33] C. Huang, Y. Chen, T. Kuo and T. Chen, Novel green-emitting $Na_2CaPO_4F:Eu^{2+}$ phosphors for near-ultraviolet white light-emitting diodes, *J. Lumin.*, 131(7), 1346–1349 (2011).

[34] Y.N. Xue, F. Xiao and Q.Y. Zhang, A red-emitting $Ca_8MgLa(PO_4)_7:Ce^{3+}$, Mn^{2+} phosphor for UV-based white LEDs application, Spectrochim. *Acta Part A*, 78(5), 1445–1448 (2011).

[35] M.G. Ha, J.H. Lee, J.S. Bae, J.P. Kim, K.S. Hong and H. Yang, Photophysical properties of highly efficient red-emitting $CaTiO_3:Eu^{3+}$ phosphors under near ultra-violet excitation, *Curr. Appl. Phys.*, 11, 1379–1383 (2011).

[36] J.S. Kim, P.E. Jeon, Y.H. Park, J.C. Choi and H.L. Park, White-light generation through ultraviolet-emitting diode and white-emitting phosphor, *Appl. Phys. Lett.*, 85, 3696–3698 (2004).

[37] Wikipedia, Dry media reaction, [http://en.wikipedia.org/wiki/Dry_media_reaction] (Last accessed, 5 Sept., 2025).

[38] A.R. West, Solid State Chemistry and Its Applications (John Wiley and Sons, Singapore) (2005).

[39] B. Gérand, G. Nowogrocki, J. Guenot and M. Figlarz, Preparative Methods in Solid State Chemistry (Academic Press, New York) (1972).

[40] C. Guo, L. Luan, C. Chen, D. Huang and Q. Su, Preparation of $Y_2O_2S:u^{3+}$ phosphors by a novel decomposition method, Mater. Lett., 62, 600–602 (2008).

[41] MT-Berlin, LED phosphors—descriptions, [http://www.mt-berlin.com/frames_cryst/descriptions/led_phosphors.htm] (Last accessed, 5 Sept., 2025).

[42] T.W. Chou, S. Mylswamy, R.S. Liu and S.Z. Chuang, Eu substitution and particle-size control of Y_2O_2S for excitation by UV LEDs, Solid State Commun., 136, 205–209 (2005).

[43] Wikipedia, Phosphor, [http://en.wikipedia.org/wiki/Phosphor] (Last accessed, 5 Sept., 2025).

[44] E.F. Schubert and J.K. Kim, Solid-state light sources getting smart, *Science,* 308, 1274–1278 (2005).

[45] Y.Q. Li, van J.E.J. Steen and J.W.H. van Krevel, Luminescence properties of red-emitting $M_2Si_5N_8:Eu^{2+}$ (M = Ca, Sr, Ba) LED conversion phosphors, J. *Alloys Compd.,* 417, 273–279 (2006).

[46] Z.L. Wang, H.B. Liang, J. Wang, M.L. Gong and Q. Su, Red-light-emitting diodes fabricated by near-UV InGaN chips with molybdate phosphors, *Appl. Phys. Lett.*, 89, 071921 (2006).

[47] N.S. Singh, R.S. Ningthoujam, N. Yaiphaba, S.D. Singh and R.K. Vatsa, Lifetime and quantum yield studies of Dy^{3+}-doped $GdVO_4$ nanoparticles: concentration and annealing effect, *J. Appl. Phys.*, 105, 064303 (2009).

[48] S. Reineke, F. Lindner, G. Schwartz, N. Seidler, K. Walzer, B. Lüssem and K. Leo, White organic light-emitting diodes with fluorescent tube efficiency, *Nature,* 459, 234–238 (2009).

[49] Y. Hu, W. Zhuang, H. Ye, S. Zhang, Y. Fang and X. Huang, Preparation and luminescent properties of $(Ca_{1-x}Sr_x)S:Eu^{2+}$ red-emitting phosphor for white LED, *J. Lumin.,* 111, 139–145 (2005).

CHAPTER 14

Flexural Phonons: A Hallmark of Freestanding 2D Materials

Satyananda Chabungbam
Department of Physics, Assam Don Bosco University, Assam, India
E-mail: csatya11@gmail.com

ABSTRACT: Flexural phonons are a critical and unique class of quasiparticles that define the properties of low-dimensional materials. Their out-of-plane vibrations, coupled with a distinct quadratic dispersion relation, lead to profound effects on thermal conductivity and mechanical behavior. As we continue to explore and utilize 2D materials for applications in electronics, thermal management, and flexible devices, a deep understanding of flexural phonons will be essential for engineering these materials with desired properties. They are not merely an interesting academic curiosity but a fundamental aspect of the physics of the 2D world.

Keywords: Quasiparticles, Out-of-plane Vibrations, Conductivity.

I. INTRODUCTION

The advent of two-dimensional (2D) materials, catalyzed by the isolation of graphene in 2004, has inaugurated a new era in condensed matter physics and materials science [Novoselov *et al.*, 2004]. These atomically thin crystals, including graphene, hexagonal boron nitride (h-BN), transition metal dichalcogenides (TMDs), and phosphorene, exhibit a remarkable confluence of physical properties not found in their bulk counterparts. Their reduced dimensionality profoundly alters their electronic band structures,

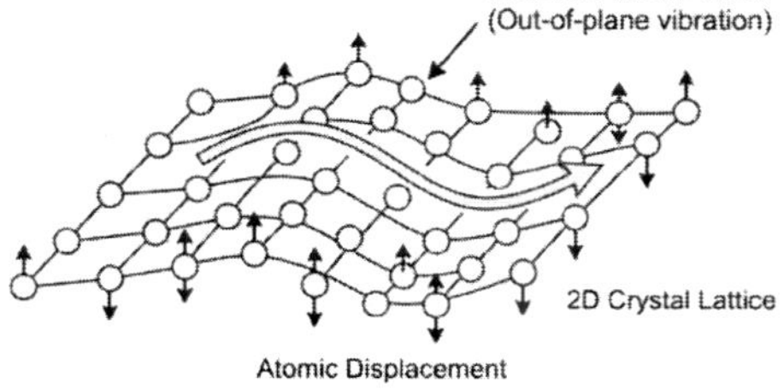

Figure 1: Flexural Phonon Modes in 2D Crystal Lattice

optical responses, and mechanical characteristics. At the heart of many of these unique behaviors lie the fundamental excitations of their crystal lattices: phonons. Phonons, the quantized modes of lattice vibration, govern thermal transport, contribute to electrical resistance through electron-phonon scattering, and determine the thermodynamic stability of a crystal.

In three-dimensional (3D) systems, the phonon spectrum is typically characterized by three acoustic modes with linear dispersion relations near the Brillouin zone center-one longitudinal and two transverse. However, in a freestanding 2D membrane, the crystal symmetry dictates a different vibrational landscape. The lattice vibrations can be decomposed into in-plane (longitudinal and transverse) and out-of-plane modes. The out-of-plane acoustic branch, known as the flexural phonon or ZA mode (for Z-axis acoustic), is a defining feature of 2D materials. Unlike its in-plane counterparts, which are governed by the strong restoring forces of bond stretching, the flexural mode corresponds to a drum-like, out-of-plane displacement of atoms as shown in Figure 1 above. The primary restoring force for these long-wavelength vibrations is the material's bending rigidity, which is significantly weaker than the in-plane tensile stiffness.

This seemingly subtle difference has a profound consequence: the dispersion relation of flexural phonons is quadratic, $\omega \propto q^2$, where ω is the phonon frequency, and q is the wave vector. This stands in sharp contrast to the linear dispersion, $\omega \propto q$, of conventional acoustic phonons. This quadratic nature leads to a very high population of low-energy flexural phonons, which fundamentally influences the material's properties. Furthermore, the out-of-plane motion of flexural phonons is subject to a mirror plane symmetry (σ_h) that imposes strict selection rules on scattering processes, further distinguishing their behavior from in-plane modes.

Indeed, the very existence of a stable, crystalline 2D material was once a subject of theoretical debate. According to the Mermin-Wagner theorem, long-wavelength fluctuations diverge logarithmically in two dimensions, which should preclude the formation of a stable crystal with long-range order [Mermin & Wagner, 1966]. The resolution to this paradox lies in the strong anharmonic coupling between the flexural and in-plane vibrational modes. This coupling suppresses the divergence of out-of-plane displacements, stabilizing the 2D lattice through the formation of microscopic ripples or corrugations [Fasolino *et al.*, 2007]. Thus, flexural phonons are not merely a feature of 2D materials; they are integral to their very existence.

This chapter provides a comprehensive review of the physics of flexural phonons. We begin by detailing their fundamental properties, including the origin of their quadratic dispersion and the unique phonon density of states they produce. We then explore the theoretical frameworks used to model these modes, from continuum elasticity theory to first-principles computations. A central theme will be the interactions of flexural phonons, both with other phonons and with electrons, which are governed by unique symmetry-imposed selection rules. Subsequently, we will discuss the macroscopic implications of these microscopic properties, focusing on their dominant role in thermal transport and mechanical stability. Finally, we will survey the applications of flexural phonon engineering in nanoelectromechanical systems (NEMS) and thermal management, and conclude by highlighting the outstanding challenges and future research directions in this vibrant field.

II. FUNDAMENTAL PROPERTIES OF FLEXURAL PHONONS

The distinct character of 2D materials is encapsulated in the unique properties of their flexural phonons. These properties arise directly from the reduced dimensionality and the nature of the restoring forces governing out-of-plane atomic displacements.

A. The Quadratic Dispersion Relation

The most defining characteristic of flexural phonons is their quadratic dispersion relation at the long-wavelength limit $q \to 0$. In a 3D crystal, displacing a plane of atoms relative to its neighbor induces a shear stress, leading to a linear restoring force and thus a linear dispersion, $\omega = v_S q$, where v_S is the speed of sound. In a 2D membrane, a long-wavelength out-of-plane displacement, or bending, does not significantly stretch the bonds. Instead, the restoring force is due to the material's resistance to curvature, known as its bending rigidity, κ.

From classical elasticity theory, the elastic energy of a thin plate due to bending is proportional to the square of its curvature. For a sinusoidal displacement profile $h(x) = A\sin(qx)$, the curvature is $\frac{\partial^2 h}{\partial x^2} = -Aq^2\sin(qx)$. The elastic energy density thus scales as $(\text{curvature})^2 \propto q^4$. The frequency of an oscillator is related to the square root of the restoring force (or the second derivative of the potential energy), leading to a frequency ω that scales with $\sqrt{q^4} = q^2$. This gives the hallmark dispersion relation for the ZA mode:

$$\omega_{ZA}(q) = \alpha q^2$$

Here, $\alpha = \kappa/\rho$, whereρ is the 2D mass density. This quadratic relationship implies that the group velocity, $v_g = \frac{\partial \omega}{\partial q} = 2\alpha q$, vanishes as $q \rightarrow 0$. This means low-energy flexural phonons are extremely slow, a property with significant consequences for thermal transport. In contrast, the in-plane longitudinal acoustic (LA) and transverse acoustic (TA) modes are governed by bond stretching and shear, respectively, and retain their linear dispersion, $\omega_{LA/TA}(q) = v_{LA/TA}q$, where $v_{LA/TA}q$ are the respective sound velocities. For graphene, for instance, $v_{LA}\approx$ 21 km/s, while the coefficient for the ZA mode is orders of magnitude smaller, reflecting the relative weakness of bending forces compared to in-plane tensile forces.

B. Phonon Density of States

The phonon density of states (DOS), g(ω), which counts the number of vibrational modes per unit frequency interval, is directly shaped by the dispersion relation. For a d-dimensional material with a dispersion of the form $\omega \propto q_v$, the DOS follows the general scaling $g(\omega) \propto \omega^{(d/v)-1}$.

For conventional acoustic phonons in 2D $(d = 2)$ with a linear dispersion $(v = 1)$, the DOS is linear in frequency:

$$g_{LA/TA}(\omega) \propto \omega^{(2/1)-1} = \omega$$

However, for flexural phonons $(d = 2)$ with their quadratic dispersion (v=2), the DOS becomes constant at low frequencies:

$$g_{ZA}(\omega) \propto \omega^{(2/2)-1} = \omega^0 = \text{constant}$$

This result is remarkable. It signifies that there is a large population of vibrational modes available at very low energies. When these modes are thermally populated according to the Bose-Einstein distribution, they can dominate the material's heat capacity at low temperatures and provide a dense phase space for scattering events. This high availability of low-energy modes is a key factor in the unique thermal and mechanical properties of 2D materials.

C. Symmetry and Selection Rules

Pristine, flat 2D materials with a hexagonal lattice (like graphene and h-BN) possess a horizontal mirror plane symmetry, denoted by the (σ_h) operation. This symmetry operation involves reflection through the plane of the material. Vibrational modes can be classified as either even or odd with respect to this operation.

In-plane modes (LA, TA): The atomic displacements lie within the plane. Reflecting through the plane leaves the displacement vectors unchanged. These modes have even parity.

Out-of-plane modes (ZA, ZO): The atomic displacements are perpendicular to the plane. Reflection reverses the direction of the displacement vectors. These modes have odd parity.

In the harmonic approximation, where phonon-phonon interactions are neglected, modes of different symmetry are orthogonal and cannot mix. This has a profound consequence for anharmonic interactions, which involve three or more phonons. For any interaction to occur, the total process must conserve crystal momentum and energy, and it must also respect the underlying symmetries of the lattice. For phonon-phonon scattering, this leads to a crucial selection rule: in any scattering process, the product of the parities of the participating phonons must be even. This implies that a scattering process must involve an even number of odd-parity (out-of-plane) phonons.

Therefore, scattering processes involving one ZA phonon, such as ZA↔LA + TA, are forbidden by this symmetry rule. Similarly, a process like ZA + ZA↔LA is also forbidden. The simplest allowed scattering channels involving flexural phonons are of the type ZA + LA↔LA or ZA + ZA↔LA + TA [Lindsay *et al.*, 2010]. This systematic prohibition of many scattering channels greatly suppresses the scattering rates for flexural phonons compared to what would be expected otherwise. This suppression is a cornerstone of our understanding of the exceptionally high thermal conductivity of materials like graphene.

III. Theoretical and Computational Modelling

Accurately describing the behavior of flexural phonons requires a hierarchy of theoretical and computational tools, ranging from continuum models that capture long-wavelength physics to atomistic simulations that incorporate quantum mechanical details.

A. Continuum Elasticity Theory

At wavelengths much larger than the interatomic spacing, a 2D material can be modelled as a continuous thin plate. The Föppl–von Kármán (FvK) theory of elasticity provides a powerful framework for this regime. The theory describes the energy of a thin plate in terms of both in-plane strain, u_{ij}, and out-of-plane displacements, h(r). The elastic free energy functional, F, can be written as:

$$F = \int d^2 r \left[\frac{1}{2}\lambda u_{ii}^2 + \mu u_{ij}^2 + \frac{1}{2}\kappa(\nabla^2 h)^2\right]$$

where λ and μ are the 2D Lamé coefficients related to the in-plane stiffness, and κ is the bending rigidity. The first two terms represent the energy from stretching and shear, while the final term represents the energy required to bend the membrane. It is this $\kappa(\nabla^2 h)^2$ term that gives rise to the quadratic dispersion. By solving the equations of motion derived from this energy functional, one can directly obtain $\omega_{ZA} = \kappa/\rho\, q^2$.

While elegant, this classical model has limitations. It treats the material as a continuum and is therefore valid only for $q \ll \pi/a$, where a is the lattice constant. Furthermore, it does not inherently account for the anharmonic coupling between in-plane and out-of-plane modes that is crucial for stabilizing the lattice. This coupling introduces nonlinear terms, such as those proportional to $(\nabla h)^2$, which are responsible for the thermal expansion and stiffening of the membrane at finite temperatures.

B. Atomistic Simulations

To capture physics at the atomic scale and to accurately model anharmonicity, researchers turn to atomistic simulations. These methods can be broadly divided into two categories:

First-Principles Calculations: These methods, primarily based on Density Functional Theory (DFT), solve the quantum mechanical equations for electrons to determine the forces on the atoms. From these forces, one can compute the harmonic force constants to generate the full phonon dispersion curves across the Brillouin zone. As an example, the phonon spectra of hexagonal Boron-Nitride generated using Quantum Espresso code [P. Giannozzi *et al* 2009, 2017, 2020] is shown in Figure 2 below. The three modes starting from 0 frequency at the Gamma point are the acoustic modes. Out of the three modes, the lowest one in red circles has quadratic nature which are known as Flexural modes, different from the above two modes which has linear character at the Gamma point. By calculating third-order (and higher) force constants, one can also quantify the strength of anharmonic phonon-phonon interactions. While highly accurate, DFT calculations are computationally expensive and are typically limited to small systems and zero temperature, although techniques like ab initio molecular dynamics can be used to study systems at finite temperatures.

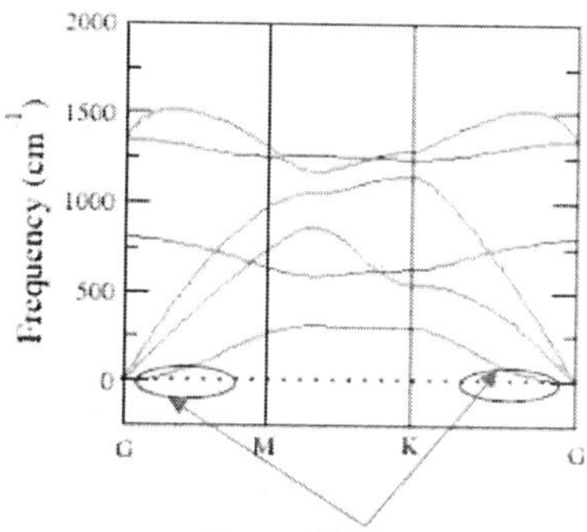

Figure 2: DFPT Simulations of Flexural Phonons of Hexagonal Boron Nitride

Molecular Dynamics (MD) with Empirical Potentials: MD simulations solve Newton's equations of motion for a large ensemble of atoms interacting via an empirical potential (or force field). Potentials like the Tersoff or REBO potentials have been developed to accurately model the covalent bonding in materials like graphene. MD simulations are computationally much cheaper than DFT, allowing for the study of millions of atoms over nanoseconds. This makes them ideal for calculating properties that depend on large-scale fluctuations and anharmonicity, such as thermal conductivity and the temperature dependence of mechanical properties. The phonon dispersion and lifetimes can be extracted from the simulations by analyzing the correlations in atomic velocities.

These atomistic methods have been instrumental in confirming the quadratic nature of the ZA dispersion and, crucially, in quantifying the anharmonic coupling strengths that govern phonon lifetimes and stabilize the 2D lattice.

C. The Role of Substrate and Tension

In most practical applications, 2D materials are not perfectly freestanding but are instead supported by a substrate or are under some degree of mechanical strain. These external factors can significantly alter the properties of flexural phonons.

Substrate Coupling: When a 2D material is placed on a substrate, van der Waals interactions create a restoring force that opposes out-of-plane motion. This effectively pins the membrane to the substrate. This introduces a constant potential energy term for displacement, which translates into an energy gap, ω_0, at the zone center ($q = 0$). The dispersion relation is modified to:

$$\omega_{ZA}^2(q) = \omega_0^2 + (\alpha q^2)^2$$

This gapping of the ZA mode can dramatically reduce its contribution to thermal transport and changes the selection rules for scattering.

Tensile Strain: Applying tensile strain to a 2D membrane pulls it taut, like a drumhead. This introduces a tension-based restoring force that is linear in displacement, similar to a vibrating string. This adds a linear term to the dispersion relation, which competes with the bending term:

$$\omega_{ZA}^2(q) = (v_T q)^2 + (\alpha q^2)^2$$

where v_T is a velocity dependent on the applied tension T. Under high tension, the linear term dominates, and the flexural mode begins to behave like a conventional transverse acoustic phonon. This "linearization" of the ZA mode stiffens the membrane, increases the phonon group velocity, and can be used as a knob to engineer the thermal and mechanical properties of the material.

IV. Interactions and Scattering Phenomena

The transport properties of a material are dictated not by the static phonon modes themselves, but by how they scatter and interact with each other and with other quasiparticles like electrons. For flexural phonons, these interactions are governed by their unique dispersion and symmetry.

A. Anharmonic Phonon-Phonon Interactions

In a real crystal, the potential energy is not perfectly harmonic, leading to interactions between phonons that limit their lifetimes and give rise to finite thermal conductivity. As discussed earlier, the (σ_h) mirror symmetry of flat 2D materials imposes a strict selection rule: an even number of out-of-plane phonons must participate in any scattering event.

This rule forbids many of the three-phonon scattering processes that would otherwise dominate thermal resistance. For instance, the decay of a high-frequency LA phonon into two lower-frequency ZA phonons (LA → ZA + ZA) is allowed, but the absorption process ZA + LA → TA is forbidden. The most important allowed scattering processes that limit the lifetime of a ZA phonon are those involving another phonon, such as:

ZA + LA ↔ LA

ZA + TA ↔ TA

ZA + ZA ↔ ZA + ZA (four-phonon process)

The phase space for these allowed processes is also restricted by the requirement to conserve both energy and momentum. Due to the quadratic dispersion of the ZA mode, it is difficult for a low-energy ZA phonon to satisfy these conditions simultaneously when scattering with high-velocity linear LA/TA modes. The combination of restrictive symmetry selection rules and phase space constraints leads to very long lifetimes for low-frequency ZA phonons [Lindsay *et al.*, 2011]. This weak scattering is the primary reason for the anomalously high thermal conductivity observed in materials like graphene and boron nitride.

B. Electron-Flexural Phonon Coupling

Electron-phonon coupling is the mechanism responsible for electrical resistance in pure metals and semiconductors and is also the basis for conventional superconductivity. In 2D materials, the coupling between electrons and flexural phonons has unique features.

The standard mechanism for electron-acoustic phonon scattering is via the deformation potential, where a longitudinal lattice distortion (compression/rarefaction) creates a local change in potential that scatters electrons. This is the primary mechanism for LA phonons. For flexural phonons, the coupling is more subtle. An out-of-plane displacement, to first order, does not change the in-plane bond lengths or area, so the standard deformation potential coupling is weak. However, coupling can arise through several second-order mechanisms:

Bond Length Modulation: In a rippled membrane, the out-of-plane displacement $h(r)$induces a local in-plane strain proportional to $(\nabla h)^2$. This strain then couples to electrons via the standard deformation potential.

Pseudo-Gauge Field: In materials like graphene with a Dirac-like electronic structure, curvature and strain can act as a pseudo-magnetic vector potential, which can effectively scatter electrons.

The electron-flexural phonon coupling is generally weaker than the coupling to in-plane optical or acoustic phonons, especially at high temperatures. However, because of the high population of thermally excited ZA phonons at low temperatures, they can become a significant contributor to electrical resistivity in this regime. Understanding this coupling is crucial for designing high-mobility electronic devices based on 2D materials.

C. Scattering by Defects and Boundaries

In addition to intrinsic phonon-phonon scattering, phonon transport is also limited by extrinsic sources like defects, impurities, and sample boundaries. Flexural phonons, being long-wavelength modes, are particularly sensitive to scattering from long-range disorder.

Point Defects: Isotopic impurities or vacancies act as point-like scattering centers. The scattering rate for this mechanism typically scales with frequency as $\tau^{-1} \propto \omega^4$ (Rayleigh scattering).

Boundaries and Roughness: At low temperatures, when intrinsic scattering is weak, phonons can travel ballistically until they hit the edge of the sample. For flexural phonons, scattering at the boundaries can be highly sensitive to the edge roughness. More importantly, large-scale features like wrinkles, folds, and microscopic ripples, which are intrinsically linked to flexural modes, act as potent scattering centers, significantly reducing phonon lifetimes and thermal conductivity. The exact nature of this scattering is complex and remains an active area of research.

V. Impact on Macroscopic Properties

The microscopic characteristics of flexural phonons; their quadratic dispersion, high density of states, and long lifetimes; manifest directly in the macroscopic thermal and mechanical properties of 2D materials.

A. Thermal Transport

Perhaps the most celebrated consequence of flexural phonon behavior is the exceptionally high intrinsic thermal conductivity (k) of materials like graphene, measured to be in excess of 3000 W/m-K at room temperature, surpassing that of diamond and copper [Balandin *et al.*, 2008]. Theoretical models attribute the majority of this heat conduction to phonons, and specifically to the ZA branch.

The contribution of a phonon mode to thermal conductivity can be estimated from the Boltzmann transport equation, which in the relaxation time approximation gives $k \propto \sum C_i v_{g,i}^2 \tau_i$, where $C_i, v_{g,i}, \tau_i$ are the specific heat, group velocity, and lifetime of mode i, respectively. For ZA phonons, the group velocity ($v_g \propto q$) is low, but this is more than compensated for by their high specific heat (due to the constant DOS) and their extremely long lifetimes (τ) arising from the symmetry-based selection rules.

A fascinating theoretical prediction emerged from this understanding: the thermal conductivity of a perfectly clean, 2D membrane should diverge with the logarithm of the sample size, L, i.e., $k \propto \log(L)$ [Lindsay *et al.*, 2010]. This divergence is a direct consequence of the ZA mode's behavior. The scattering rate for the lowest-frequency ZA phonons is so low that their mean free path can exceed the sample size. As the sample gets larger, these ultra-long-mean-free-path phonons contribute more to the overall conductivity, leading to the divergence. While true logarithmic divergence is an idealization, this behavior underscores the quasi-ballistic nature of flexural phonon transport and has been partially observed in simulations and experiments, where the measured thermal conductivity shows a strong dependence on sample dimensions.

B. Mechanical Stability and Rippling

As mentioned in the introduction, flexural phonons are central to the thermodynamic stability of 2D crystals. The Mermin-Wagner theorem states that in 2D, the mean-squared atomic displacement, $\langle h^2 \rangle$, caused by long-wavelength thermal fluctuations should diverge logarithmically with system size, destroying long-range crystalline order.

The solution lies in the anharmonic coupling between out-of-plane and in-plane modes. As the amplitude of a flexural mode increases, it pulls atoms closer together in-plane, effectively creating a local tensile strain. This strain stiffens the membrane, which in turn suppresses the amplitude of the flexural fluctuations. This nonlinear feedback mechanism dynamically stabilizes the lattice by converting large out-of-plane displacements into microscopic, static ripples. These intrinsic ripples, with typical heights of a few angstroms and wavelengths of several nanometres, have been directly observed using scanning tunnelling microscopy. While they ensure the stability of the 2D crystal, these ripples are a form of structural disorder that can scatter electrons and phonons, impacting electronic mobility and thermal conductivity.

VI. Applications and Engineering

A deep understanding of flexural phonons is not just of fundamental interest; it also opens avenues for engineering novel devices and functionalities.

A. Nanoelectromechanical Systems (NEMS)

The unique mechanical properties of 2D materials make them ideal candidates for NEMS resonators. An atomically thin membrane of graphene or MoS_2 can be suspended over a trench and actuated electrostatically to vibrate at very high frequencies (MHz to GHz range). The fundamental resonant frequency is determined by the tension and mass of the membrane, and its motion is a manifestation of the lowest-energy flexural mode. The performance of a resonator is characterized by its quality factor (Q-factor), which measures the ratio of stored energy to energy dissipated per cycle. High Q-factors are desirable for applications in sensing and signal processing. The energy dissipation in these resonators is largely due to anharmonic phonon-phonon interactions. The restricted scattering channels for flexural phonons can contribute to very high Q-factors in 2D material resonators, making them promising for ultra-sensitive mass sensors, clocks, and filters [Barton *et al.*, 2011].

B. Thermal Management and Thermoelectrics

The ability to manipulate thermal conductivity is critical for thermal management in microelectronics and for improving the efficiency of thermoelectric devices. Flexural phonons offer several knobs for tuning heat flow in 2D materials.

Strain Engineering: As discussed, applying tensile strain linearizes the ZA dispersion, increasing their group velocity and altering their scattering rates. This can be used to either increase or decrease thermal conductivity, depending on the interplay between velocity and lifetime changes.

Substrate Engineering: Placing a 2D material on different substrates can control the degree of ZA mode suppression. Strongly interacting substrates can "freeze out" the flexural modes, drastically reducing the out-of-plane thermal conductivity.

Phononic Crystals: Patterning a 2D material with a periodic array of holes or pillars can create a "phononic crystal" with a tailored phonon band structure. This can be used to open bandgaps for flexural phonons, effectively blocking their transport and creating thermal insulators or filters.

For thermoelectric applications, the goal is to achieve a high figure of merit, $ZT = (S^2 \sigma/k)T$, which requires high electrical conductivity (σ) and low thermal conductivity (k). The fact that electrons couple relatively weakly to flexural phonons, while flexural phonons dominate thermal conductivity, suggests a pathway to decouple electrical and thermal transport. By selectively scattering flexural phonons (e.g., using nanopatterning), one could potentially reduce k without significantly harming σ, thereby enhancing the thermoelectric efficiency.

VII. CHALLENGES AND FUTURE OUTLOOK

Despite significant progress, the field of flexural phonons is rich with open questions and exciting future prospects.

A. Theoretical and Computational Hurdles

Accurately modelling the anharmonic interactions that govern flexural phonon lifetimes remains a formidable challenge. While DFT can provide accurate third-order force constants, including higher-order (e.g., four-phonon) processes, which are thought to be important in 2D materials, is computationally prohibitive. Developing more efficient and accurate computational schemes that can bridge the gap between atomistic simulations and device-scale models is a critical need. Furthermore, a complete quantum mechanical treatment of nuclear motion, going beyond the classical approximations of MD, may be necessary to fully capture the behavior of flexural modes at low temperatures.

B. Experimental Challenges

Directly probing the properties of low-energy flexural phonons is experimentally difficult. Techniques like inelastic neutron or X-ray scattering, which are standard for measuring phonon dispersions in bulk crystals, suffer from low signal-to-noise ratios when applied to a single atomic layer. While electron energy loss spectroscopy (EELS) in a transmission electron microscope has had some success, it has limited momentum resolution. Developing new experimental techniques to directly measure the dispersion, and more importantly, the lifetimes of flexural phonons as a function of wavevector and temperature, would provide crucial data to validate and refine theoretical models.

C. Emerging Frontiers

The role of flexural phonons in more complex 2D systems is a rapidly growing area of research.

Moiré Superlattices: In twisted bilayer graphene and other Moiré heterostructures, the periodic Moiré potential can fold the phonon bands, create new "phono-Moiré" modes and alter the flexural phonon properties. Understanding how this affects thermal transport and electron-phonon coupling in these strongly correlated electronic systems is a key frontier.

2D Magnets and Excitonic Systems: The coupling between flexural phonons and other quasiparticles is a largely unexplored territory. How do out-of-plane vibrations interact with magnons (spin waves) in 2D magnets? How do they affect the properties of excitons (electron-hole pairs) in semiconducting TMDs? These interactions could lead to novel hybrid excitations and device concepts.

Novel Materials: The vast library of 2D materials continues to expand. Investigating the properties of flexural phonons in anisotropic materials like black phosphorus, or in more exotic materials like 2D ferroelectrics and topological insulators, promises to reveal new physics and functionalities.

VIII. Conclusion

Flexural phonons are not a mere curiosity of two-dimensional systems; they are a central feature that dictates their fundamental properties. Their unique quadratic dispersion relation, a direct consequence of the physics of thin plates, leads to a high population of low-energy modes. This, combined with symmetry selection rules that severely restrict their scattering, results in exceptionally long phonon lifetimes. This collection of properties makes flexural phonons the dominant heat carriers in many 2D materials, explains the anomalous dependence of thermal conductivity on system size, and is intrinsically linked to the rippling that ensures the stability of the 2D crystalline state itself.

The ongoing exploration of flexural phonons continues to push the boundaries of our understanding of thermal, mechanical, and electronic phenomena in the 2D limit. The ability to engineer their properties through strain, substrates, and nanostructuring provides a powerful toolkit for designing next-generation nanoelectronic devices, thermal management solutions, and NEMS. As we venture into the rich landscape of 2D heterostructures and novel quantum materials, the physics of flexural phonons will undoubtedly continue to be a source of fascinating fundamental discoveries and technological innovation.

References

[1] A.A. Balandin, S. Ghosh, W. Bao, I. Calizo, D. Teweldebrhan, F. Miao and C.N. Lau, Superior thermal conductivity of single-layer graphene, Nano Letters, 8(3), 902–907 (2008).

[2] R.A. Barton, B. Ilic, A.M. van der Zande, W.S. Whitney, P.L. McEuen, J.M. Parpia and H.G. Craighead, High size-dependent quality factor in graphene drum resonators, Nano Letters, 11(3), 1232–1236 (2011).

[3] A. Fasolino, J.H. Los and M.I. Katsnelson, Intrinsic ripples in graphene, *Nature Materials*, 6(11), 858–861 (2007).

[4] L. Lindsay, D.A. Broido and N. Mingo, Flexural phonons and thermal transport in graphene, Physical Review B, 82(11), 115427 (2010).

[5] L. Lindsay, D.A. Broido and N. Mingo, Flexural phonons and thermal transport in multilayer graphene and graphite, *Physical Review B*, 83(23), 235428 (2011).

[6] N.D. Mermin and H. Wagner, Absence of ferromagnetism or antiferrom agnetism in one- or two-dimensional isotropic Heisenberg models, *Physical Review Letters*, 17(22), 1133 (1966).

[7] K.S. Novoselov, A.K. Geim, S.V. Morozov, D. Jiang, Y. Zhang, S.V. Dubonos, I.V Grigorieva and A.A. Firsov, Electric field effect in atomically thin carbon films, *Science*, 306(5696), 666-669 (2004).

[8] P. Giannozzi *et al.*, *J. Phys.:Condens. Matter*, 21 395502 (2009).

[9] P. Giannozzi *et al.*, *J. Phys.: Condens. Matter*, 29 465901 (2017).

[10] P. Giannozzi *et al.*, *J. Chem. Phys.* 152 154105 (2020).

[11] http://www.quantum-espresso.org

CHAPTER 15

Phenomenology of Joglekar Window Function for Thin Film Memristor

Jyoti Prasad Roy Choudhury[1,2*], Barnali Pathak[2], Nayan Mani Nath[3] and Bikash Borah[1]

[1]Department of Physics, Assam Don Bosco University, Guwahati, India

[2]Department of Physics, B.H. College, Howly, Barpeta, India

[3]Department of Physics, Barnagar College, Sorbhog, Barpeta, India

*E-mail: jyotirc62@gmail.com

ABSTRACT: With an efficient simulation technique, we have studied the memristive behavior of the ZnO based thin films using Joglekar window function. Simulated results using the model are compared with experimental values and observed considerable consistency in this regard for a particular set of parameters. In addition, we comment on several issues such as boundary effect, boundary lock, and scalability in this regard.

Keywords: Memristor, Window Function, Boundary Lock Phenomenon.

I. INTRODUCTION

Nowadays, human civilization has been relying upon and consistently generating huge amount of data and hence it is in great need of an advanced technology for their treatment as well as preservation. Memristor, as the most promising technology to deal with massive amount of data have attracted much attention in recent researches and significant progresses have been observed in this regard [1-3]. Due to presence of a unique characteristic (pinched hysteresis loop) in voltage and current relation, memristor has been expected to behave as an efficient memory element. In order to establish a direct link between charge and magnetic flux, in 1971 Leon Chua proposed a theory to build their linkage [4] and later in 2008 Scientists from HP lab developed a practical device and called as Memristor [5]. This device has the properties of both a memory element and a resistor, hence wisely named as memristor.

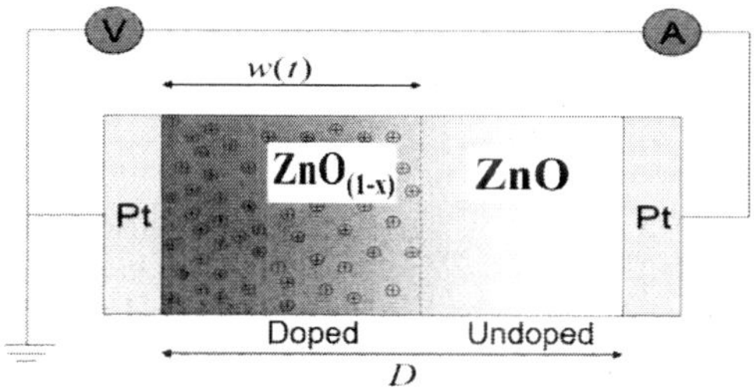

Figure 1: Model of ZnO based Memristor

Most memristors adopt a metal/insulator/metal structure and for metal electrodes, TiO_2 have been a popular choice. To understand memristors and memristive behavior there exist a wide variety of theoretical as well as phenomenological models, predictions of which are being analyzed in comparison with actual experimental data in order to verify the validity of the approaches as well as their applicability. In this report we present our results on memristive ZnO based thin films by means of an effective theoretical model and efficient simulation technique.

In this study we present our results on memristive ZnO based thin films by means of an effective theoretical model and efficient simulation technique. ZnO is chosen as the active material due to its wide band gap, good thermal and chemical stability, and excellent compatibility with low-cost thin-film fabrication techniques. Moreover, the presence of intrinsic oxygen vacancies in ZnO plays a crucial role in resistive switching behavior, making it a suitable material for memristive applications. We have also compared our results with experimental values and observe a very good agreement.

II. Theory

In the study of memristors, available mathematical approaches are linear, non-linear, exponential, Simmons Tunnel Barrier, and TEAM models which incorporate several window functions such as Joglekar, Biolek, Strukov, Prodromakis, Zha etc. In 2008, Stanley Williams and his group at HP Labs realized the memristor in device form and that is called under consideration as shown in Figure 1. The governing equations of such a memristive system can be written as follows [5, 7]

$$V(t) = M(t)i(t) \qquad \dots (1)$$

With the total resistance of doped and un-doped region or total memristance,

$$M(t) = R_{ON}x + R_{OFF}(1 - x) \qquad \dots (2)$$

$$M(t) = R_{OFF} - (R_{OFF} - R_{ON})x = R_{OFF} - \Delta_R x$$

Here $x = \frac{w(t)}{D}$ is the state variable with w and D, the width of the doped layer and total thickness of the memristor respectively. The state derivative $\frac{dx}{dt}$ is the moving speed of boundary between the doped and undoped layer which is characterized by

$$\frac{dx}{dt} = \eta i(t) \quad \text{... (3)}$$

Where $\eta = \mu_v \frac{R_{ON}}{D^2}$, R_{ON} and R_{OFF} are the values of resistance for $x = 0$ and $x = 1$ respectively and μv the dopant mobility. Above equation is a linear equation because velocity of the width w(t) is linearly proportional to the current and hence this model is known as linear drift model. However, it is observed that in nano-scale region a small change in voltage causes enormous electric field which makes the ionic transport nonlinear and therefore, linear model is inadequate in this regard. Hence a nonlinear model is needed to achieve nonlinear characteristics to the above model. In order to incorporate non-linear characteristics Joglekar and Wolf developed a modified model by means of introducing a nonlinear function, known as window function f(x)with the above equation and which in turn becomes

$$\frac{dx}{dt} = \eta i(t) f(x) \quad \text{... (4)}$$

A window function is a mathematical function that has zero valued outside some chosen interval. When another function multiplied by the window function, the product is always zero valued outside the interval. There exist several window functions in literature for memristor.

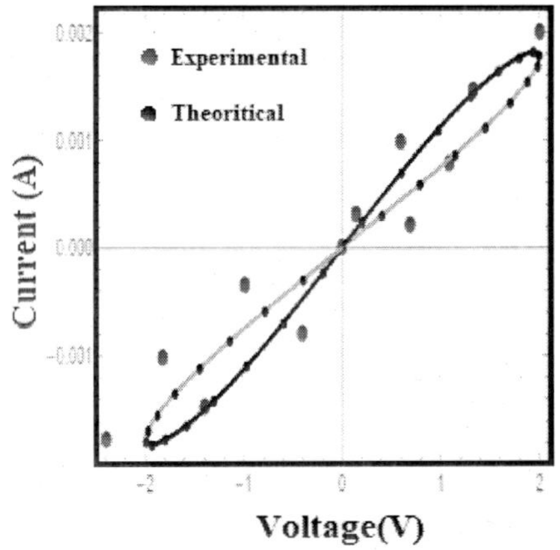

Figure 2: I(t)–V(t) Characteristic Curve of 30 nm ZnO Based thin Film Memristor

One of the important and interesting window functions is Joglekar window function and it is given by [7].

$$f(x) = 1 - (2x - 1)^{2p} \qquad \dots (5)$$

Here p is a positive integer and it controls the linearity of the model.

We have solved eq. (4) using the Joglekar windows function,

$$f(x) = F[W(t), p] = 1 - (2W - 1)^{2p} \qquad \dots (6)$$

and obtain current from eq. (4) as

$$i(t) = \frac{W'(t)}{\eta . F[W(t), p]} \qquad \dots (7)$$

Using this expression for current in eq. (1), we get

$$V(t) = (R_{ON} - \Delta_R W(t)) * \frac{W'(t)}{\eta . F[W(t), p]} \qquad \dots (8)$$

Further, the function W(t) is determined by solving the differential eq. (4) as

$$W'(t) = \eta F[W(t), p] \frac{V(t)}{M(t)} = \frac{V_0 \sin(2\pi t/T)}{R_{OFF} - \Delta_R W(t)} \qquad \dots (9)$$

Thus, the state equation used in Mathematica is obtain by solving the equations for I(t), V(t) and W′(t) respectively.

III. Results and Discussion

Using a simulation code developed in MATHEMATICA, we have obtained the I(t)−V(t) characteristic curve for ZnO based memristor of thickness 30nm and the corresponding results are shown in Figure 2. Here the small black dots along with black and green curves are our simulation results. Alongside our results, we have plotted experimental data taken from [9] and they are represented by blue points [see [9] for details]. Our results shown here are in accord with the parameters used in Ref. [9]. Comparative analysis reflects considerable consistency between theoretical and experimental results as well as their memristive behavior, the hysteresis loop, which in turn verifies the prediction capability and experimental evidence of memristive property in ZnO thin films.

Although Joglekar window function is capable of predicting memristive nature and solves nonlinear drift behavior but it has several drawbacks. It is observed that for $x = 0$ and $x = 1$ the state derivative $\frac{dx}{dt}$ vanishes i.e., when boundary arrives at either side of the device, there is no external stimulus to change the state. Therefore, for two extreme cases, the memristance of the memristor no longer varies. This is known as boundary lock phenomenon.

Moreover, it is observed that for the various value of p, f(x) value does not scale up and scale down. Hence, the nonlinear I-V curves remain the same for this window function. This problem is known as scalability. There exist several window functions to resolve these issues and all of them succeeded more or less.

IV. Conclusion

We obtain the I(t)−V(t) characteristics of ZnO based thin film memristor using Joglekar window function and performed phenomenological analysis with the experimental data taken from [9]. We observe satisfactory result in this regard. Simultaneously, we have discussed the limitations such as boundary lock, scalability etc. of Joglekar window function. Although the window functions succeeded in prediction to some extent but not yet complete. Therefore, obtaining a complete and efficient window function is an interesting problem in recent research.

References

[1] F. Caravelli and J.P. Carbajal, Memristors for the Curious Outsiders, Technologies, 2018, 6, 1–43.

[2] E. Apollos, Memristor Theory and Mathematical Modelling, International Journal of Computer Applications (0975–8887), 2019, 178, 1–8.

[3] L. Chua, Everything You Wish to Know About Memristors but Are Afraid to Ask, Radioengineering, 24, 319–368.

[4] L. Chua, Memristor: The missing circuit element, IEEE Trans. Circuit Theory, 1971, 18, 507–519.

[5] D.B. Strukov, D.S. Snider, D.R. Stewart and R.S. Williams, The missing memristor found, *Nature,* 2008, 453, 80–83.

[6] L. Chua, If it's pinched it's a memristor, *Semicond. Sci. Technol,* 2014, 29, 104001 (1–42).

[7] C. Hu, Q. Wang, S. Bai, M. Xu, D. He, D. Lyu and J. Qi, The effect of oxygen vacancy on switching mechanism of ZnO resistive switching memory, *Appl. Phys. Let.*, 2017, 110, 073501(1–4).

[8] Y.N. Jogleker and J.S. Wolf, The elusive memristor: properties of basic electrical circuits, *Eur. J. Phys.*, 2009, 30, 661–675.

[9] F. Gul and H. Efeoglu, ZnO and ZnO_{1-x} based thin film memristors: The effects of oxygen deficiency and thickness in resistive switching behavior, *Ceramics International*, 2017, 43, 10770–10775.

CHAPTER 16

Quaternary Borocarbide Superconductors: Structure, Properties and Applications

Santosh Kumar Barik[1], Basanta Kumar Sahoo[2*], Nilanchala Patel[3], Jitendra Sahoo[4] and Bibekananda Panda[5]

[1]PG Department of Physics, Utkal University, Bhubaneswar, Odisha, India
[2]PG Department of Physics, Govt. Autonomous College, Angul, Odisha, India
[3]PG Department of Chemistry, Govt. Autonomous College, Angul, Odisha, India
[4]Regional Directorate of Education, Bhubaneswar, Odisha, India
[5]SDVM Nilakanthanagar, Berhampur, Odisha, India
*E-mail: sahoo75.cmt@gmail.com

ABSTRACT: The category of materials that combines structural complexity and richness with a multiplicity of physical phenomena is represented by the rare-earth quaternary borocarbide superconductors, which have the general formula RNi_2B_2C. *Since their discovery, they have attracted significant attention because of their comparatively high critical temperatures (up to 23 K) and the unusual coexistence of magnetism and superconductivity. These compounds crystallize in a tetragonal lattice structure in which* Ni_2B_2 *layers are separated by RC sheets, giving rise to a quasi-two-dimensional electronic structure. This layered arrangement not only enhances superconductivity but also promotes anisotropic pairing. One of the most interesting peculiarities related to borocarbides is the relationship between magnetism and superconductivity. Nonmagnetic members, e.g.* YNi_2B_2C, $LuNi_2B_2C$, *have relatively high transition temperatures and exhibit unusual gap symmetries, possibly comprising mixtures of s- and d-wave components. Magnetic variants such as* $HoNi_2B_2C$ *and* $ErNi_2B_2C$ *are, in contrast, rare cases of antiferromagnetism and superconductivity to form a platform with which to study competing quantum orders. The other* RNi_2B_2C *(R = Tm, Dy, Ho, Er) are mainly magnetic superconductor systems in which both AFM and SC orders are present. The large upper critical fields, large critical current densities and strong flux-pinning properties make them attractive for use in superconducting magnets, cryogenic detectors and microwave devices. Moreover, other common applications of the borocarbide superconductors include spintronics, fusion reactors, smart grids and power transmission, and numerous others. Furthermore, their stability in structure and moderate transition temperatures make it possible to incorporate them into future hybrid superconducting systems that can be used at the liquid-helium temperature. Future progress in this field will rely on both experimental research and theoretical modelling to focus on the processes that underlie their atypical characteristics and to advance their use in future technologies of next-generation superconducting devices.*

Keywords: Borocarbide Superconductors, High-T_c Materials, Antiferromagnetism, De Gennes Factor.

I. Introduction

Quaternary borocarbides are alternatively referred to as RNi_2B_2C systems [1]. The label R appears for a rare-earth element, such as Y, Lu, Er, Ho, or Dy, in the quaternary borocarbide superconductors. They represent a unique class of intermetallic compounds bridging conventional and unconventional superconductivity [2, 3]. Since their discovery in the mid-1990s, borocarbides have attracted considerable scientific attention. They are particularly interesting because they support both superconductivity and magnetic order simultaneously in the same structure [4, 5]. Unlike conventional superconductors, where magnetism usually suppresses superconductivity, borocarbides maintain a delicate balance between antiferromagnetic order (AFM) and superconductivity. This unique feature allows detailed studies of how localized magnetic moments influence the formation of Cooper pairs [2, 4, 5].

In structure, borocarbides crystallise in the tetragonal I4/mmm space group, where rare-earth atoms and nickel–boron sheets are arranged in layers and carbon fills the interstitial sites [6]. This layered structure exhibits considerable anisotropy and provides methods to tune both the electronic and magnetic properties by chemical substitution, external pressure, or doping [7]. RNi_2B_2C compounds are characterized as quaternary superconductors because they consist of four different elements with SC Critical temperatures (T_C) that vary from approximately 6 K to 16 K, depending on which rare-earth element is present.

These borocarbides are situated at the scientific level between high-temperature cuprate superconductors and conventional phonon-based systems [2, 7]. The highly ordered crystal structure, intermediate electron–phonon coupling, and the presence of strongly localized 4f magnetic moments make borocarbides ideal for studying several exciting features. These include anisotropic superconducting (SC) gaps, multiband pairing, and the competition between magnetism and superconductivity. All of these are important aspects of condensed matter research today [1, 3, 8].

II. TYPES, FAMILIES, AND STRUCTURE

The borocarbide family of superconductors can be systematically categorised based on the magnetic properties of the incorporated rare-earth element (R). Generally, they are classified into two main groups: nonmagnetic borocarbides such as YNi_2B_2C and $LuNi_2B_2C$ [9], and magnetic borocarbides including $ErNi_2B_2C$, $HoNi_2B_2C$, and $DyNi_2B_2C$ [5]. The nonmagnetic compounds serve as ideal model systems for investigating intrinsic SC mechanisms without the influence of magnetism. On the other hand, the magnetic counterparts offer an arena to study the complex relationship between magnetism and superconductivity [2, 5].

Crystallographically, these materials have a body-centred tetragonal structure with the space group I4/mmm [11]. They have a crystal lattice that is composed of alternate layers of R-C and Ni_2B_2, giving a clear quasi-two-dimensional electronic structure [12]. In this lattice, nickel atoms form square planar networks, which are similar to the CuO_2 planes in the cuprate superconductors. The interstitial sites are occupied by boron and carbon atoms, with strong covalent bonding, which increases the stability and rigidity of the lattice. Sheets of conductive Ni_2B_2 separate the layers of rare-earth ions. This separation facilitates electronic anisotropy and hybridized orbital interactions [12, 13].

The unit cell is characterized by lattice parameters near a = 3.5 Å and c = 10.5 Å, with small deviations in the c/a ratios over the rare-earth series. These deviations affect the electronic bandwidth, strength of hybridisation and hence, T_c and Néel temperature (T_N) [10]. Structural stability arises from

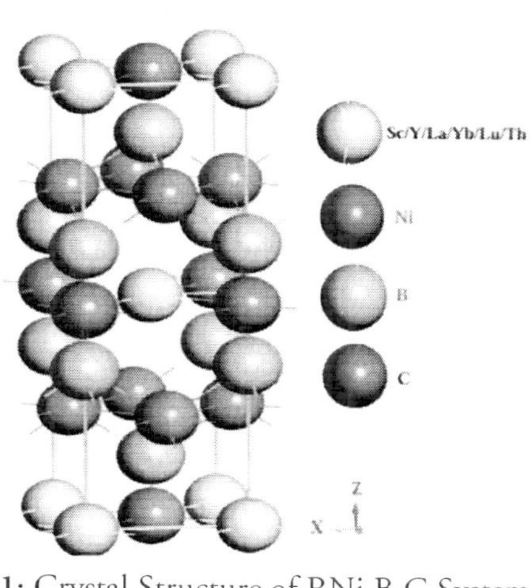

Figure 1: Crystal Structure of RNi_2B_2C Systems [13]

strong hybridization between Ni 3d and B 2p orbitals. Additionally, carbon donates electrons to the system, enhancing the density of states (DOS) at the Fermi level. Various band structure calculations reveal a complex multiband electronic nature with many sheets in the Fermi surface [10, 12], which are significant in producing the observed anisotropic superconducting gap and associated effects.

III. Electronic and Phonon Properties

A. Electronic Structure and Multiband Fermi Surface Topology

The quaternary borocarbide superconductors (RNi_2B_2C) provide a uniquely rich platform for studying the interplay among superconductivity, magnetism, and anisotropic electron–phonon coupling within a single family of intermetallic compounds [14]. Their electronic structure is highly multiband in nature, arising from substantial contributions of Ni 3d, B 2p, and C 2p orbitals [15]. Density functional theory calculations reveal that the Fermi surface is composed of a quasi-two-dimensional cylindrical sheet coexisting with multiple three-dimensional pockets [16]. These unusual topological features result in a highly anisotropic superconducting gap. They also create multiple momentum-dependent electron-phonon interactions, especially in regions connected by favourable nesting vectors [17].

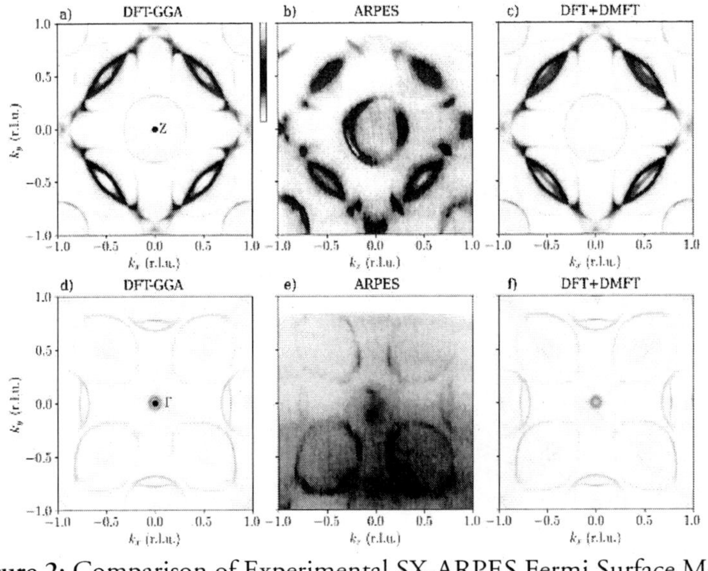

Figure 2: Comparison of Experimental SX-ARPES Fermi Surface Maps to One-Step Model of Photoemission [18]

B. Experimental Evidence from ARPES, Quantum Oscillations, and Neutron Scattering

ARPES and the de Haas-van Alphen effect have been used to experimentally verify the existence of these distinct components of the Fermi surface. These techniques have shown that the superconducting energy gap is extremely anisotropic and very low-energy phonon branches show a strong association with portions of the Fermi surface [19]. Specifically, the interactions between Ni–B stretching and B–C shear modes have a strong momentum dependence that is related directly to the multiband electronic structure [20]. ARPES studies also indicate that electrons in the quasi-two-dimensional sheet are more strongly coupled to low-energy phonons than they are in the three-dimensional pockets.

Inelastic neutron scattering measurements support these observations by showing anomalous softening of several phonon branches. Both transverse acoustic (TA) and low-energy optical modes are strongly softened near the phonon wave-vectors corresponding to characteristic nesting vectors of the quasi-two-dimensional band [21]. This softening is closely associated with superconducting behaviour, indicating that the interplay between Fermi surface nesting and phonon dispersion plays a key role in stabilizing superconductivity. Spatial mapping further reveals that regions of the Fermi surface contributing most to nesting exhibit the largest phonon anomalies [22, 23].

C. Momentum-Dependent Electron–Phonon Coupling and Isotope Effects

Further evidence is provided by the studies of boron and carbon phonon modes in isotope-effect research on the essential role thereof. Partial isotope sensitivity is observed upon isotope substitution, giving a reduced change in the critical temperature despite being less sensitive, resulting in a reduced change in the critical temperature [24]. The reduced isotope exponent can be explained by multiband effects and the presence of magnetic rare-earth ions. These ions influence phonon softening and modify the electron–phonon interaction via magnetic exchange. The low-energy phonon spectrum is also affected by AFM or SDW ordering in alloys such as $HoNi_2B_2C$ and $ErNi_2B_2C$. Notably, gaps appear in Fermi surface regions connected through nesting. Consequently, magnetic order competes with superconductivity, either by breaking Cooper pairs or altering the phonon spectrum [25, 26].

D. Interplay of Magnetic Ordering and Anisotropic Superconductivity

Out of all the non-magnetic materials, YNi_2B_2C and $LuNi_2B_2C$ exhibit a non-conventional, anisotropic superconducting behaviour. Expanding ARPES results on deep gap minima or similar nodal-like behaviour along some chosen crystallographic directions seem to indicate an intricate, possibly anisotropic s-wave or multigap s-wave superconducting state. Complementary specific heat and thermal conductivity measurements indicate that at low temperatures, there exist power-law behaviour, consistent with nodal quasiparticles. These observations indicate the multiband nature of superconductivity in this family, in which interband strength and anisotropic gap structures have a large role in low temperature behaviour [27, 28, 29].

IV. Interplay between Antiferromagnetism and Superconductivity

A. Coexistence of Superconductivity and Antiferromagnetism in RNi_2B_2C

The quaternary rare-earth nickel borocarbides (RNi_2B_2C) are unique intermetallic compounds. They challenge the conventional belief that superconductivity (SC) and long-range antiferromagnetic (AFM) order cannot coexist within the same crystal lattice [30]. This anomaly arises from the simultaneous presence of superconductivity and antiferromagnetism on the same crystal lattice after a comprehensive study of these unique intermetallics demonstrated that SC and AFM can co-exist [5]. This implies that SC is associated with the localized 4f ions that release electrons upon the application of a magnetic field, while AFM is associated with itinerant 3d and 2p conduction electrons that would generate Cooper Pairing [12, 31]. Each of the above classes of intermetallics exhibits a typical SC energy gap value of between 1.5 and 2.8 meV and a magnetic exchange energy of between 3 and 12 meV [32]. Consequently, the SC and AFM energy scales are partially overlapping, and accordingly provide separate energy fields for the formation of SC and AFM under specific conditions. Moreover, the phonon energy spectra (10 to 25 meV) obtained via inelastic neutron scattering and Raman spectra indicate that superconductivity is primarily phonon-mediated [18].

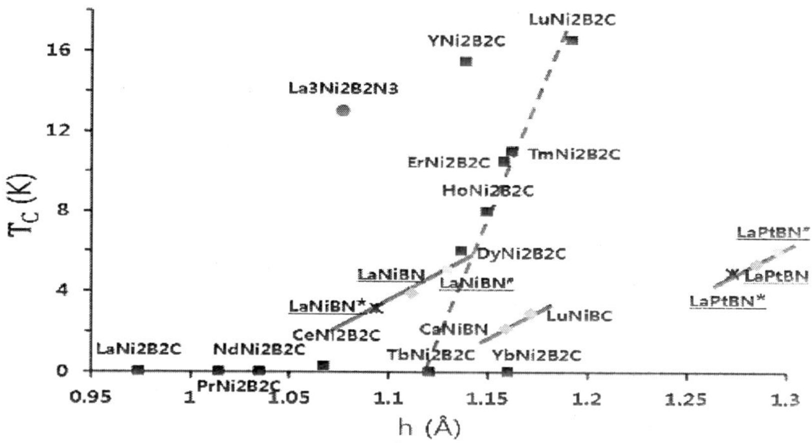

Figure 3: The Relation between the Superconducting Critical Temperature T_c and the B Height h in the M-B Layer in Borocarbides and Boronitrides [33]

B. Variation of Superconductivity and Magnetic Ordering Across Rare-Earth Elements

The SC transition temperature varies significantly across quaternary borocarbides, depending on the rare-earth element present (Y, Lu, Er, Ho, Dy, or Tm). Here are the measured values of T_c for the different compounds: YNi_2B_2C ($T_c \approx 15.6K$), $LuNi_2B_2C$ ($T_c \approx 16.6K$), $ErNi_2B_2C$ ($T_c \approx 10.5K$), $HoNi_2B_2C$ ($T_c \approx 8.5K$), $DyNi_2B_2C$ ($T_c \approx 6.2K$), $TmNi_2B_2C$ ($T_c \approx 11K$) [34, 35]. Using Neutron diffraction, there is clear evidence showing that AFM order will begin to emerge around or below T_c. In contrast, $HoNi_2B_2C$ has a

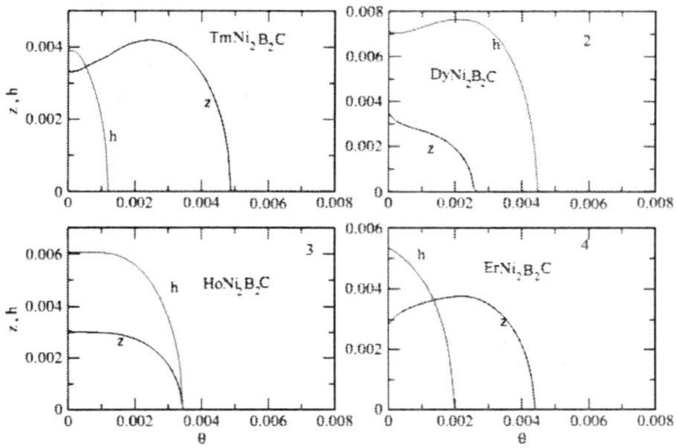

Figure 4: Dimensionless SC Gap (z) and AFM Gap (h) *vs.* Temperature Parameter (θ) for Magnetic Borocarbide Superconductors RNi_2B_2C (R=Tm, Dy, Ho, Er) [38]

T_c of about 8.5 K and develops long-range AFM order around $T_N \approx 5$ K. Its magnetic moment exhibits both commensurate and incommensurate spatial modulations along the a-axis. In the case of $ErNi_2B_2C$, $T_N \approx 6$ K, $ErNi_2B_2C$ has re-entrant superconductivity, where the SC is partially suppressed near T_N, only to re-enter at lower temperatures after strong interactions of conduction electrons and ordered 4f magnetic moments [36, 37].

Using techniques such as muon spin rotation (μSR) and Nuclear Magnetic Resonance spectroscopy (NMR), researchers have shown that the magnetic field from the AFM structure does not fully suppress the superconducting order parameter. Experimental data confirm theoretical predictions that magnetic and superconducting interactions occur at different intrinsic energy scales, as described by Ginzburg-Landau and Eliashberg theory. Instead, the localized f moments remain separated in the real space [39].

C. Systematic Trends, de Gennes Scaling, and Resulting T–H Phase Diagram Complexity

Systematic trends throughout the series of rare earths reveal a close relationship between T_N, T_c, and the de Gennes factor, i.e. $(gJ - 1)^2J(J + 1)$. This factor is proportional to the strength of the exchange interaction I and density of states $N(\varepsilon_F)$ at Fermi level. Moving from Tm to Dy, due to increase in de Gennes factor (and thus increase in magnetic order), it is found that T_c decreases while T_N remains constant (due to increased pair-breaking); in agreement with both theoretical and experimental results [40, 41]. For example, while $T_N \approx 10.5$ K of $DyNi_2B_2C$ exceeds its $T_c \approx 6.2$ K, the opposite is true for $TmNi_2B_2C$ where $T_N \approx 1.5$ K and $T_c \approx 11$ K.

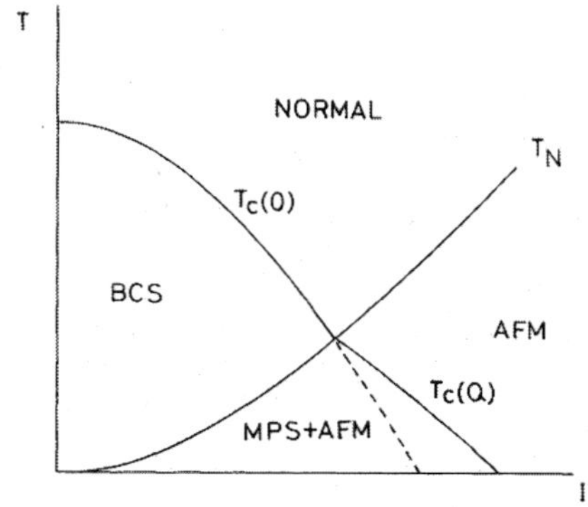

Figure 5: Schematic Phase Diagram of Antiferromagnetism (AFM) and Superconductivity (BCS and MPS) as a Function of the Exchange Integral I [42]

The various kinds of competition, coexistence and mutual interaction of S/C behaviour in RNi_2B_2C yield fairly complex T/H phase diagrams consisting of many AFM transition points, vortex order shifts and re-entrant possibilities – all of which can be seen in $HoNi_2B_2C$ and $ErNi_2B_2C$.Therefore, these four-component borocarbides provide an important platform for understanding competing electronic orderings in the systems of correlated electrons [43].

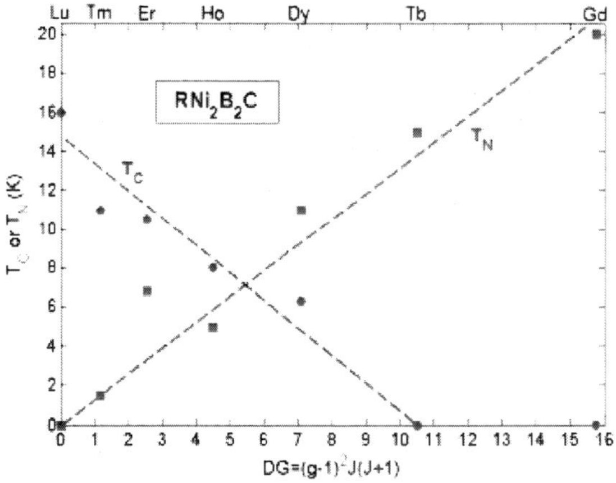

Figure 6: T_c and T_N Scale Roughly Linearly with the de Gennes Factor (DG) [26]

5. Experimental and Theoretical Models

An in-depth investigation of RNi_2B_2C is done through a close combination of theory and experiment. Experimental measurements in techniques like neutron diffraction, X-ray photoelectron spectroscopy (XPS), ARPES, scanning tunnelling microscopy (STM) and magnetisation measurements have significantly helped in gaining a better insight into the structural, electronic and magnetic attributes of these materials [44, 45] .

An important method for determining the crystal as well as the magnetic structure of the RNi_2B_2C compounds is neutron scattering. Elastic diffraction has shown long-range magnetic order in Ho, Er and Dy-based RNi_2B_2C compounds, and inelastic scattering the spin-wave excitations that show strong interaction of localized 4f moments with the itinerant electrons [5]. Scanning tunnelling spectroscopy (STS) enable you to view the SC gap directly by itself and indicates the anisotropic and multiband character of superconductivity in YNi_2B_2C and $LuNi_2B_2C$ [7, 26]. ARPES measurements allow mapping of the momentum-resolved SC gap. They also help determine how Fermi surface nesting contributes to electron-phonon interactions. The presence of SC and magnetic phases is confirmed with the help of measurements of magnetisation, resistivity and specific heat [2, 46].

Theoretical approaches employ ab initio techniques like DFT and DFT+U to study the connection between Ni 3d and R 4f states to reproduce the Fermi surfaces and calculate the electron–phonon coupling that corresponds to experimental critical temperatures. Apart from mean-field approaches, tight-binding, Hubbard-type, and Ginzburg–Landau-based models address electron correlations, gap anisotropy, and the conflict between superconducting and magnetic order parameters, providing a holistic explanation for coexistence and competition in these materials [44, 46].

VI. Applications in Industrial & Sustainable Development

A. Mechanical, Electrical, and Processing Advantages of RNi_2B_2C Superconductors

Although their transition temperatures are moderate, the RNi_2B_2C compounds yield significant benefits both in the technological and sustainability aspects. Their high level of metallic conductivity, ductile mechanical behaviour and high intermetallic bond strength enable them to overcome most limitations of high-Tc cuprates. The natural metallicity of the borocarbides permits standard machining, steady thermal expansion, and compatibility with various thin-film deposition techniques. These allow them to be used in devices whose structural reliability and mechanical integrity are as important as actually being in the superconducting state [47, 48, 49].

B. Thin-Film Growth, Device Integration, and High-Frequency Performance

High-quality thin films of YNi_2B_2C and $LuNi_2B_2C$ have been prepared by pulsed-laser deposition, sputtering and molecular-beam epitaxy [50]. These films sustain high critical current densities without any problems in their critical and clearly defined energy gaps and superior thermal cycling resistance [51]. As an alternative to cuprates, where grain-boundary weakening, anisotropic d-wave pairing and intrinsic brittleness are problematic, the borocarbides are more stable than cuprates in forming cryogenic wires, microwave devices, and superconducting circuits. Their clean-limit operation and low intrinsic disorder reduce microwave losses. As a result, they can be used in resonators, detectors, Josephson junctions, and hybrid superconducting-semiconductor systems [52].

They also have a low microwave surface resistance, which is another useful property because it improves performance in higher frequency applications. Compared to films consisting of niobium, which under some circumstances can be used to achieve higher dissipation, borocarbide films have lower losses because they have lower defect density and are well bonded with metallic bonding [53]. For this reason, they are now considered for cryogenic electronics, precision sensing, and components requiring reproducible electromagnetic performance [20].

C. Magnetocaloric Functionality and Multifunctional Cryogenic Applications

Borocarbides containing magnetic rare-earth elements have additional functionality due to their magnetocaloric properties. They undergo significant entropy changes in magnetic fields, which is advantageous for cryogenic magnetic refrigeration systems, particularly near the Neel temperature. They exhibit excellent thermal stability, good heat capacity and excellent corrosion resistance, which increases the life cycle of cooling systems and superconducting devices. The interplay between superconducting and magnetocaloric behavior enables new strategies for designing materials with both cooling and superconducting functions. This could lead to compact, energy-efficient architectures for quantum sensing and cryogenic networks.

D. Sustainability Considerations, Operational Constraints, and Emerging Engineering Challenges

In terms of sustainability, borocarbides offer long-term material efficiency. Boron, nickel and carbon are relatively common elements, and there is an increasing trend in the recycling of rare-earth. Their extended periods of operation and non-disposal to environmental damage decrease material turnover, which is in favour of the principles of the circular economy. They also facilitate energy transportation and compact superconducting motors that provide low-loss energy transport and help in increasing energy efficiency in cryogenic power and sensing technologies [54, 55].

Although they have these benefits, there are some limitations to their scope of use. Their Tc values require liquid-helium cooling, limiting their competitiveness in temperature ranges where MgB_2 or iron-based superconductors operate more efficiently. Additionally, in magnetic borocarbides, the interplay between magnetic ordering and superconductivity must be carefully engineered to minimize magnetic pair-breaking. However, progress in thin-film growth, nanoscale materials engineering, and composite systems continues to enhance their performance and expand their applicability in industrial and scientific systems [56, 57, 58, 59].

VII. Future Scope and Challenges

Although considerable advancements have been done, quaternary borocarbides superconductors are not easy to study. They also provide inspiration on possible future directions of research. Their superconducting and magnetic properties need to be better studied and enhanced in order to realize their power as a technology. Their applicability in potential application is decreased by the low T_c. Chemical substitution, strain, and external pressure methods have resulted in rather modest Tc gains [60, 61]. Stronger overlap on electronic band structure with lattice compression may be the main cause of moderate increments in pressure studies [5, 62].

Although coupling between antiferromagnetism and superconductivity is well-documented, its microscopic mechanisms remain poorly understood. Contemporary techniques such as neutron scattering, resonant X-ray spectroscopy, and ultrafast pump-probe spectroscopy are needed for further investigation. Resolving complex phase transitions and re-entrant behaviour of spin-orbitally coupled weakly-interacting intervening phase transitions should motivate further theoretical and experimental developments. Future studies could explore quaternary iterations involving mixed rare-earth compounds, such as (Ho, Er)Ni_2B_2C, to achieve tunable magneto-superconducting properties. The future directions of combining experimental resolution and advanced modelling might provide answers to the hybrid superconducting-magnetic systems [62].

VIII. Conclusion

Quaternary borocarbide superconductors constitute a remarkable class of materials that promises to be a good research topic in the realm of superconductivity, magnetism, and structural integrity. After being discovered, other materials such as RNi_2B_2C (R = Y, Lu, Ho, Er, Dy, Tm) have become important systems with which to study the phenomena of interplay and antagonism between antiferromagnetism and superconductivity. The crystal structure of these materials, which is a layer of metallic and covalent planes, forms a complex game of competing localized magnetic moments and the creation and binding of pairs of electrons. Extensive experimental evidence and theoretical modelling show that superconductivity in borocarbides is primarily mediated by phonons and is highly anisotropic and is multiband in nature.

From an application perspective, it can be anticipated that borocarbides may be of great application in industries as well as in sustainable application. Their appealing mechanical qualities, corrosion behaviour and the ability to show dependable superconductivity at cryogenic temperatures make them attractive to be used in coatings of superconductors, cryoelectronics and magnetocaloric refrigeration innovations. They are also recyclable and can be easily composed; another desirable trait that suits the sustainability-based material development. It is expected that the combination of highly advanced experiment tools with theoretical models will only increase their technological importance and benefit to condensed matter studies.

ACKNOWLEDGEMENT

The authors express their heartfelt appreciation to the Director, Institute of Physics, Bhubaneswar, Odisha, India, for granting them access to research facilities throughout their visit. Additionally, Mr. Santosh Kumar Barik wishes to convey his gratitude to the Department of Higher Education, Government of Odisha, India for conferring the Research Fellowship under the MRIP scheme.

REFERENCES

[1] M. Kawamura, R. Akashi and S. Tsuneyuki (2017). Anisotropic superconducting gaps in YNi_2B_2C: A first-principles investigation. *Phys. Rev. B,* 95(5), 054506.

[2] S.L. Bud'ko and P.C. Canfield (2006). Magnetism and superconductivity in rare earth–nickel–borocarbides. *C. R. Phys.,* 7(1), 56–67.

[3] R. Nagarajan (1997). Superconductivity and magnetism in quaternary borocarbides. *J. Low Temp. Phys.*, 107(5), 517–526.

[4] T. Baba, T. Yokoya, S. Tsuda, T. Kiss, T. Shimojima, K. Ishizaka ... and S. Shin (2008). Bulk electronic structure of the antiferromagnetic superconducting phase in $ErNi_2B_2C$. *Phys. Rev. Lett.,* 100(1), 017003.

[5] S. Gundogdu, J.P. Clancy, G. Xu, Y. Zhao, P.A. Dube, T.C. Karalar, ... and M. Ramazanoglu (2020). Magnetic order and competition with superconductivity in (Ho-Er)Ni_2B_2C. *Mater. Res. Express,* 7(11), 116002.

[6] G.M. Schmiedeshoff, De Boer, C., M.V. Tompkins, W.P. Beyermann, A.H. Lacerda, J.L. Smith and P.C. Canfield (2000). Superconductivity (and magnetism) in the nickel borocarbides. *J. Supercond.*, 13(5), 847–853.

[7] H.M. Tütüncü, E. Karaca and G.P. Srivastava (2017). Electron–phonon interaction and superconductivity in the borocarbide superconductor. *Philos. Mag.*, 97(29), 2669–2688.

[8] H. Kawano-Furukawa, Y. Ishida, F. Yano, R. Nagatomo, A. Noda, T. Nagata, ... and H. Takeya (2010). Creation of vortices by ferromagnetic order in $ErNi_2B_2C$. *Physica C*, 470, S716–S718.

[9] H. Takagi, M. Nohara and R.J. Cava (1997). Borocarbide superconductors: Materials and physical properties. Physica B, 237, 292–295.

[10] L. Liu, X. Wu, R. Wang, Z. Hu, Y. Jiang and D. Liu (2017). First principles study on structure stability and mechanical properties of YNi_2B_2C and $LuNi_2B_2C$ under pressure. Crystals, 7(6), 173.

[11] K. Izawa, K. Kamata, Y. Nakajima, Y. Matsuda, T. Watanabe, M. Nohara, ... and K. Maki (2002). Gap function with point nodes in borocarbide superconductor YNi_2B_2C. *Phys. Rev. Lett.*, 89(13), 137006.

[12] E. Pellegrin, C.T. Chen, G. Meigs, R.J. Cava, J.J. Krajewski and W.F. Peck (1995). Jr. Electronic structure of RNi_2B_2C (R = rare earth) from x-ray-absorption spectroscopy. Phys. Rev. B, 51(22), 16159.

[13] M.A. Rahman, N. Algethami, M.A. Khatun, M.Y.H. Khan, M.N.M. Nobin, T. Akhtar and A. Basak (2025). Elastic, electronic, thermal, and superconducting properties of Ni-based borocarbides MNi_2B_2C (M = Sc, Y, La, Yb, Lu, Th) through DFT replication. *Braz. J. Phys.*, 55(6), 254.

[14] R.J. Cava, H. Takagi, H.W. Zandbergen, J.J. Krajewski, W.F. Peck, Jr., T. Siegrist, ... and S. Uchida (1994). Superconductivity in the quaternary intermetallic compounds $LnNi_2B_2C$. *Nature*, 367(6460), 252–253.

[15] S.B. Dugdale, M.A. Alam, I. Wilkinson, R.J. Hughes, I.R. Fisher, P.C. Canfield ... and G. Santi (1999). Nesting properties and anisotropy of the Fermi surface of $LuNi_2B_2C$. *Phys. Rev. Lett.*, 83(23), 4824.

[16] S.L. Drechsler, H. Rosner, S.V. Shulga, I. Opahle, H. Eschrig, J. Freudenberger ... and A. Gladun (2001). A brief comparison of superconductivity in borocarbides and cuprates. *Physica C*, 364, 31–36.

[17] S.V. Shulga, S.L. Drechsler, G. Fuchs, K.H. Müller, K. Winzer, M. Heinecke and K. Krug (1998). Upper critical field peculiarities of superconducting YNi_2B_2C and $LuNi_2B_2C$. *Phys. Rev. Lett.*, 80(8), 1730.

[18] A. Pulkkinen, G. Kremer, V.N. Strocov, F. Weber, J. Minár and C. Monney (2025). Revealing electronic correlations in YNi_2B_2C using photoemission spectroscopy. *Commun. Phys.*, 8(1), 256.

[19] E. Boaknin, R.W. Hill, C. Proust, C. Lupien, L. Taillefer and P.C. Canfield (2001). Highly anisotropic gap function in borocarbide superconductor $LuNi_2B_2C$. *Phys. Rev. Lett.*, 87(23), 237001.

[20] F. Weber, L. Pintschovius, W. Reichardt, R. Heid, K.P. Bohnen, A. Kreyssig, ... and K. Hradil (2014). Phonons and electron-phonon coupling in YNi_2B_2C. *Phys. Rev. B*, 89(10), 104503.

[21] H. Kawano, H. Yoshizawa, H. Takeya and K. Kadowaki (1996). Anomalous phonon scattering below T_c in YNi_2B_2C. *Phys. Rev. Lett.*, 77(22), 4628.

[22] H.M. Tütüncü, H.Y. Uzunok, E. Karaca, G.P. Srivastava, S. Özer and Ş. Uğur (2015). Ab initio investigation of BCS-type superconductivity in $LuNi_2B_2C$-type superconductors. *Phys. Rev. B*, 92(5), 054510.

[23] B. Bergk, V. Petzold, H. Rosner, S.L. Drechsler, M. Bartkowiak, O. Ignatchik, ... and J. Wosnitza (2008). Anisotropic multiband many-body interactions in $LuNi_2B_2C$. *Phys. Rev. Lett.*, 100(25), 257004.

[24] K.O. Cheon, I.R. Fisher and P.C. Canfield (1999). Boron isotope effect in single-crystal YNi_2B_2C and $LuNi_2B_2C$ superconductors. *Physica C*, 312(1–2), 35–39.

[25] H. Hodovanets, S. Ran, P.C. Canfield and S.L. Bud'ko (2013). Boron isotope effect in single crystals of superconductor. *Philos. Mag.*, 93(14), 1748–1754.

[26] K.H. Müller and V.N. Narozhnyi (2001). Interaction of superconductivity and magnetism in borocarbide superconductors. *Rep. Prog. Phys.*, 64(8), 943.

[27] C.L. Huang, J.Y. Lin, C.P. Sun, T.K. Lee, J.D. Kim, E.M. Choi, ... and H.D. Yang (2006). Comparative analysis of specific heat of YNi_2B_2C using nodal and two-gap models. *Phys. Rev. B*, 73(1), 012502.

[28] T. Baba, T. Yokoya, S. Tsuda, T. Watanabe, M. Nohara, H.T.T. Oguchi and S. Shin (2008). Relation between superconducting gap minima and nesting vector in YNi_2B_2C. arXiv:0812.0406 [cond-mat.supr-con].

[29] C.L. Huang, C.P. Sun, J.Y. Lin, H.J. Kim, E.M. Choi, S.I. Lee and H.D. Yang (2006). Specific heat of YNi_2B_2C analyzed using two-gap model. *AIP Conf. Proc.*, 850(1), 685–686.

[30] L.C. Gupta (2006). Superconductivity and magnetism and their interplay in quaternary borocarbides RNi_2B_2C. *Adv. Phys.*, 55(7–8), 691–798.

[31] X. Lu, W.K. Park, S. Yeo, K.H. Oh, S.I. Lee, S.L. Bud'ko, ... and L.H. Greene (2011). Superconducting order parameter in nonmagnetic borocarbides RNi_2B_2C (R = Y, Lu) probed by point-contact Andreev reflection spectroscopy. Phys. Rev. B, 83(10), 104519.

[32] Y.G. Naidyuk, G. Behr, N.L. Bobrov, V.N. Chernobay, S.L. Drechsler, G. Fuchs, ... and I.K. Yanson (2009). The superconducting gap behavior in the antiferromagnetic nickel-borocarbide compounds RNi_2B_2C (R = Dy, Ho, Er, Tm) studied by point-contact spectroscopy. J. Phys.: Conf. Ser., 150(5), 052178.

[33] M.C. Jung, C.J. Kang, B.I. Min and K.W. Lee (2013). Electronic structures and phonon spectra in boronitride superconductors LaMBN (M = Ni, Pt). Phys. Rev. B, 87(14), 144509.

[34] S.L. Drechsler, H. Rosner, S.V. Shulga, G. Fuchs, H. von Lips, J. Freudenberger, ... and K. Koepernik (1999). Electronic structure and superconductivity of nonmagnetic transition metal borocarbides. J. Low Temp. Phys., 117(5), 1617–1621.

[35] P. Allenspach and U. Gasser (2000). Rare earth magnetism in high-temperature and borocarbide superconductors. J. Alloys Compd., 311(1), 1–10.

[36] S. Gabani, K. Flachbart, K. Siemensmeyer and T. Mori (2020). Magnetism and superconductivity of rare earth borides. J. Alloys Compd., 821, 153201.

[37] B.K. Sahoo and B.N. Panda (2015). Competition of superconductivity and antiferromagnetism in RNi_2B_2C (R = Tm, Dy, Ho, Er). Physica C, 513, 35–38.

[38] B.K. Sahoo, B.N. Panda and F.T. Dias (2014). Study of de-Gennes factor and effect of external magnetic field on SC and AFM in borocarbide superconductors. IJATES, 2(11), 445–450.

[39] B.K. Cho, P.C. Canfield, L.L. Miller, D.C. Johnston, W.P. Beyermann and A. Yatskar (1995). Magnetism and superconductivity in single-crystal $ErNi_2B_2C$. Phys. Rev. B, 52(5), 3684.

[40] B. Johansson, L. Nordström, O. Eriksson and M.S.S. Brooks (1991). Magnetism in rare-earth metals and rare-earth intermetallic compounds. Phys. Scr., 1991(T39), 100.

[41] B.K. Sahoo, S. Pradhan, S. Das, S.K. Barik, P. Patnaik, R.K. Bhuyan and B.N. Panda (2022). Study of effective de-Gennes factor and effect of doping on superconductivity in rare-earth nickel borocarbide superconductors. Mater. Today: Proc., 66, 3315–3318.

[42] K. Machida, K. Nokura and T. Matsubara (1980). New pairing state and partial destruction of pairing in antiferromagnetic superconductors. Phys. Rev. Lett., 44(12), 821.

[43] M. ElMassalami, R.M. Saeed, C.M. Chaves, H. Takeya, M. Doerr, H. Michor and M. Rotter (2010). Specific heat investigation on the magnetic phase diagrams of RNi_2B_2C (R = Gd, Er). J. Magn. Magn. Mater., 322(5), 523–529.

[44] V. Christiansson and P. Werner (2024). Quaternary borocarbides: A testbed for DFT for superconductor. Phys. Rev. B, 109(18), L180505.

[45] P. Kurzhals, G. Kremer, T. Jaouen, C.W. Nicholson, R. Heid, P. Nagel, ... and F. Weber (2022). Electron-momentum dependence of electron-phonon coupling underlies dramatic phonon renormalization in YNi_2B_2C. Nat. Commun., 13(1), 228.

[46] P. Martinez-Samper, H. Suderow, S. Vieira, J.P. Brison, N. Luchier, P. Lejay and P.C. Canfield (2003). Phonon-mediated anisotropic superconductivity in the Y and Lu nickel borocarbides. Phys. Rev. B, 67(1), 014526.

[47] C.R. Tomassetti, G.P. Kafle, E.T. Marcial, E.R. Margine and A.N. Kolmogorov (2024) (2024). Prospect of high-temperature superconductivity in layered metal borocarbides. *J. Mater. Chem. C*, 12(13), 4870–4884.

[48] G. Fuchs, K.H. Müller, J. Freudenberger, K. Nenkov, S.L. Drechsler, S.V. Shulga, ... and P. Gegenwart (2002). Influence of disorder on superconductivity in non-magnetic rare-earth nickel borocarbides. Pramana, 58(5), 791–797.

[49] A. Gurevich (2023). Tuning microwave losses in superconducting resonators. Supercond. Sci. Technol., 36(6), 063002.

[50] T. Niemeier, R. Hühne, A. Köhler, G. Behr, L. Schultz and B. Holzapfel (2010). Biaxially textured $LuNi_2B_2C$ thin films on MgO single crystals. J. Alloys Compd., 507(2), 345–349.

[51] O.E. Kvitnitskaya, Y.G. Naidyuk, I.K. Yanson, T. Niemeier, G. Fuchs, B. Holzapfel and L. Schultz (2010). Point-contact study of the $LuNi_2B_2C$ borocarbide superconducting film. Supercond. Sci. Technol., 23(11), 115001.

[52] T. Honma, S. Sato, K. Sato, M. Watanabe, A. Saito, K. Koike, ... and S. Ohshima (2013). Microwave surface resistance of YBCO superconducting thin films under high DC magnetic field. Physica C, 484, 46–48.

[53] M.D. Lan, J.C. Chang, K.T. Lu, C.Y. Lee, H.Y. Shih and G.Y. Jeng (2002). Upper critical field of borocarbide superconductors. IEEE Trans. Appl. Supercond., 11(1), 3607–3610.

[54] M.R. Koblischka, D.M. Koblischka, A. Koblischka-Veneva, E.A.D.S. Duran, R. Izquierdo and R. Zadorosny (2025). Superconductivity and the sustainable development goals (SDGs): A challenge for researchers in superconductivity. Sustainability, 17(9), 4068.

[55] N.H. Andersen, N. Bay, J.C. Grivel, P. Hedegård, D. McMorrow, S. Mørup, ... and B. La (2003). Superconductivity and magnetism: Materials properties and developments. Risø Int. Symp. Mater. Sci., Risø Natl. Lab.

[56] Y.Y. Hsu, H.C. Chiang and H.C. Ku (1998). Superconductivity in the new quaternary ruthenium borocarbide Y–Ru–B–C system. J. Appl. Phys., 83(11), 6789–6791.

[57] N.T.M. Duc, H. Srikanth and M.H. Phan (2025). Low-dimensional magnetocaloric materials for energy-efficient magnetic refrigeration: Does size matter? STAM, 26(1), 2546287.

[58] M.A. Rahman, A. Khatun, D.C. Roy, M. Chowdhury and A. Irfan (2025). DFT simulation to study the mechanical, electronic, thermal and super conducting properties of borocarbides materials AX_2B_2C where A = Y, La, Th and X = Pd, Pt. Sci. Rep., 15(1), 30033.

[59] J.Y. Law, L.M. Moreno-Ramírez, Á. Díaz-García, and V. Franco (2023). Current perspective in magnetocaloric materials research. J. Appl. Phys., 133(4).

[60] R. Falconi, A. Durán and R. Escudero (2001). High-pressure effects in single crystals: RNi_2B_2C (R = Dy, Pr). Phys. Rev. B, 65(2), 024505.

[61] J. Freudenberger, K.H. Müller, G. Fuchs, M. Wolf, K. Nenkov and Schultz (2008). Pair weakening in mixed rare-earth nickel borocarbides $(R, R')Ni_2B_2C$. arXiv:0811.1694 [cond-mat.supr-con].

[62] H.D. Liu, W.Y. Zhang, Z.G. Fu, B.T. Wang, H.Y. Lu, H.J. Song, ... and P. Zhang (2025). Multi-gap and high-Tc superconductivity in metal-atom-free borocarbides: Effects of dimensional confinement and strain engineering. arXiv:2507.03284 [cond-mat.supr-con].

CHAPTER 17

Majorana Materials: Topological Superconductors for Resource-Efficient Quantum Computing

Priyanuj Rajbongshi
Department of Physics, Indian Institute of Technology Tirupati, Tirupati, India
E-mail: ph24d004@iittp.ac.in

ABSTRACT: Majorana fermions are an unusual class of particles: they are their own antiparticles. In certain superconducting materials, they appear in pairs, usually at the ends of a tiny wire or inside a vortex. Because the two modes live far apart, information can be stored nonlocally and is harder for noise to spoil, as they are topologically protected. This chapter explains the concept using the well-known Kitaev chain, where Majorana zero modes can be observed at the edges of the chain. We tour the main material platforms: semiconductor-superconductor nanowires, magnetic topological-insulator hybrids, atomic chains on superconductors, and iron-based superconductors. Readers will learn what good evidence looks like: hard superconducting gaps, zero-bias features that are stable and quantised, and Josephson effects with 4π periodicity and what can mimic these signals. Finally, we ask why Majoranas matter for society. If they allow quantum computers with less error-correction cost, we may need fewer devices and, more importantly, less cryogenic power per useful qubit. That means smaller energy use and materials demand.

Keywords: Majorana Zero Mode, Topological Superconductor, Kitaev Chain, Sustainable Quantum Computing.

I. INTRODUCTION

Quantum computing requires qubits that are not easily perturbed by their environment. Majorana zero modes (MZMs), quasiparticles that are their own antiparticles, have emerged as a promising route to realize topologically protected qubits. In superconducting systems supporting MZMs, information can be stored nonlocally across a pair of spatially separated Majorana modes, making the qubit inherently robust against local noise [1]. This prospect of fault-tolerant quantum computation was first outlined by Kitaev in 2003 [2], building on the exotic non-Abelian statistics of Majorana fermions predicted in certain condensed matter settings. Realizing MZMs in the lab has since become a major goal in condensed matter physics and quantum engineering.

This chapter provides a self-contained overview of Majorana materials systems that can host Majorana zero modes and their implications for resource-efficient quantum computing. We begin with the theoretical foundation via the Kitaev chain model [1], a minimal toy model of a 1D topological superconductor that yields Majorana modes at its ends [3, 4]. We then survey the leading experimental platforms for MZMs, including semiconductor nanowires [5] with superconductors, topological insulator-superconductor hybrids, chains of magnetic atoms on superconductors, and certain iron-based superconductors [6]. For each platform, we discuss the experimental signatures observed, such as zero-bias conductance peaks and fractional Josephson effects [7, 8], and how to distinguish true Majorana signals from confounding effects like disorder or trivial Andreev bound states. In the final sections, we reflect on the broader impact. If topological Majorana qubits can reduce the overhead of error correction, they could enable quantum computers that consume less cryogenic power and use fewer devices, an important consideration for sustainability. We also highlight concerns about the critical materials involved (for example, indium in nanowires and tellurium in topological insulators) and consider more earth-abundant alternatives. The overall aim is to bridge the science of Majorana materials with their potential role in building a sustainable quantum technology for the future.

A. Kitaev Chain Model: Majorana Zero Modes in One Dimension

The simplest model that captures Majorana zero modes is the Kitaev chain [1], introduced by Alexei Kitaev. It consists of a one-dimensional chain of sites with spinless fermions, including nearest-neighbor hopping and p-wave superconducting pairing. The Hamiltonian can be written (in second quantization) as:

$$H = -\mu\sum_{j=1}^{N}\left(c_j^{\dagger}c_j - \frac{1}{2}\right) + \sum_{j=1}^{N-1}\left[-t\left(c_{j+1}^{\dagger}c_j + c_j^{\dagger}c_{j+1}\right) + \Delta\left(c_{j+1}^{\dagger}c_j^{\dagger} + c_jc_{j+1}\right)\right]$$

where $c_j^{\dagger}$, c_j are creation and annihilation operators on site *j*, *t* is the hopping amplitude, μ is the chemical potential, and Δ is the superconducting pairing potential (taken real for simplicity). By introducing Majorana operators-effectively splitting each ordinary fermion into two Majorana fermions (real and imaginary parts) on each site. Kitaev showed that the chain can exhibit two distinct phases: a trivial superconducting phase and a topological superconducting phase. In the topological phase, the chain ends host unpaired Majorana zero modes that are absent in the trivial phase, as shown in Figure 1, the two near-zero eigenvalues correspond to Majorana zero modes. Crucially, these Majorana end modes are zero-energy solutions that do not appear in the bulk spectrum. They result in a two-fold degenerate ground state, corresponding to whether the pair of Majoranas is occupied or empty, despite costing zero energy to change this occupancy. In other words, a single nonlocal fermionic state is formed by combining the two end Majoranas, and it can be either filled or unfilled without breaking any pairing–this is the hallmark of a Majorana zero mode.

A key insight from the Kitaev chain solution is that the existence of Majorana modes is topologically protected. Figure 2 demonstrates the exponentially decaying probability density corresponding to the eigenstates of the zero-energy Majorana modes, which are quite robust and indicate the topological nature of the system. The system can only switch between trivial and topological phases by closing the superconducting gap (a quantum phase transition) - for the 1D chain, this happens when $|\mu| = 2t$ which marks the phase boundary. In the topological regime (e.g. μ in a suitable range less than the critical value), the end Majoranas persist as long as the gap remains open and are insensitive to small perturbations that do not close the gap. This robustness is what makes them attractive for quantum computing:

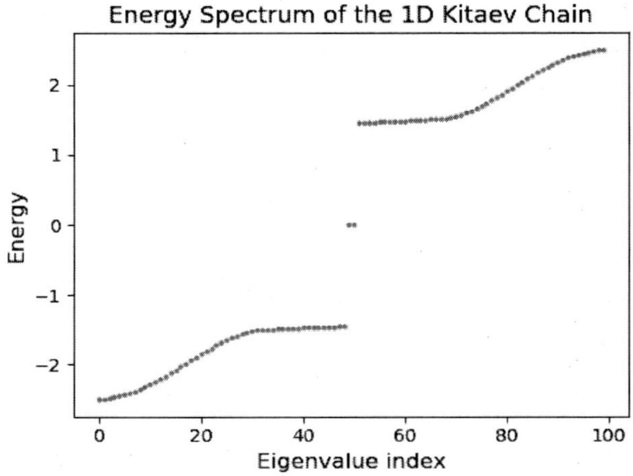

Figure 1: Energy spectrum of the Kitaev chain Hamiltonian with parameters ($N = 50, t = 1.0, \mu = 0.5, \Delta = 0.8$). The two nearly zero-energy eigenvalues at the center of the spectrum correspond to the Majorana zero modes.

operations on the degenerate ground state formed by these distant Majoranas can, in principle, be done in a way that depends only on topological properties, not on fine local details, thereby naturally resisting decoherence. In practice, a physical Kitaev chain can be engineered using more conventional ingredients, as we will see, for instance, a semiconductor nanowire with strong spin–orbit coupling in proximity to an s-wave superconductor effectively realizes a Kitaev-like p-wave paired chain in the presence of a magnetic field. The Kitaev chain thus serves as the conceptual foundation for Majorana materials, illustrating how topology and superconductivity can produce zero-energy Majorana bound states at the ends of a 1D system [9]. We next explore how this concept has been implemented in real material platforms.

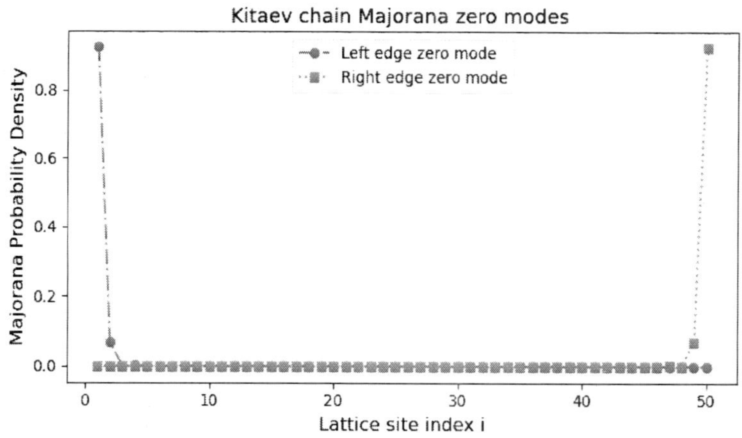

Figure 2: Spatial profile of the two Majorana zeromodes in a one-dimensional Kitaev chain with open boundary conditions ($N = 50, t = 1.0, \mu = 0.5, \Delta = 0.8$). The left and right zero-energy states are localized exponentially at opposite ends of the chain, as indicated by their respective probability densities. This localization is a hallmark of the topological superconducting phase ($|\mu| < 2t$).

II. Experimental Platforms for Majorana Modes

A. Semiconductor-Superconductor Nanowires

One of the most studied platforms for Majorana zero modes is the semiconductor nanowire with induced superconductivity. This approach was proposed in 2010 by two independent theories [7, 8] (often referred to as the Lutchyn-Oreg model. The setup uses a nanowire made of a semiconductor with strong spin orbit coupling (e.g. indium arsenide (InAs) or indium antimonide (InSb)), placed in contact with a conventional s-wave superconductor (like Al or Nb). With an applied longitudinal magnetic field, the system effectively becomes a spinless p-wave superconducting wire, in other words, a physical realization of a Kitaev chain. Above a certain magnetic field, the wire enters a topological superconducting phase with Majorana zero modes localized at its ends.

Experimental evidence: In 2012, Mourik *et al.* [7] reported the first signature consistent with Majorana modes in such a device. They observed a zero-bias conductance peak (ZBP) in tunnelling spectroscopy at the end of an InSb nanowire when a magnetic field was applied: a peak appearing at zero voltage bias in the differential conductance, suggestive of a zero-energy state at the wire's end. The zero-bias peak emerged above a critical magnetic field and was robust over a range of field and gate voltage settings, as expected for a Majorana zero mode, while disappearing or moving away from zero in the presumed trivial phase.

Challenges: Despite these promising results, semiconductor nanowire platforms face challenges. One major issue is disorder. Early experiments showed a "soft gap" and ZBPs that could sometimes be explained by disorder or quantum dot states in the wire rather than true topological Majorana modes. It is now understood that in a disordered or non-uniform wire, trivial bound states can localize near the wire end and also produce zero bias peaks, especially if coupled to a quantum dot or in the presence of smooth potential variations. Distinguishing such trivial zero-energy states from genuine Majoranas requires careful analysis of how the zero bias feature behaves under parameter changes.

B. Topological Insulator–Superconductor Hybrids

Another platform for Majorana zero modes builds on the physics of topological insulators (TIs). A topological insulator like Bi_2Se_3 or Bi_2Te_3 has conducting Dirac surface states with spin-momentum locking. Fu and Kane (2008) theorized that if an s-wave superconductor is put in contact with a TI surface, the induced pairing on the helical surface state effectively creates a 2D topological superconductor that can host Majorana modes in its vortex cores [10]. In essence, a vortex in the TI's superconducting surface binds a zero-energy Majorana state at its core, provided time-reversal symmetry is broken.

Experimental evidence: Pioneering experiments in this direction used hybrid structures like Bi_2Se_3 or Bi_2Te_3 thin films on superconductors. In a 2015 study, Xu *et al.* grew Bi_2Te_3 on the surface of $NbSe_2$ (a layered s-wave superconductor) and used scanning tunnelling microscopy (STM) to probe vortex cores [10]. They reported observing a zero-bias peak in the tunneling conductance at the vortex centre on the TI surface, which was interpreted as evidence of a Majorana zero mode.

Another variation of this platform involves magnetism on the topological insulator surface. If one introduces a magnetic gap on a TI surface (for instance, by doping the TI with magnetic impurities or interfacing it with a ferromagnet), one can create 1D channels or domain walls that host chiral Majorana modes when superconductivity is present. A notable (and controversial) experiment in 2017 by He *et al.* involved a quantum anomalous Hall (QAH) insulator, essentially a thin TI film doped with magnetic atoms to produce a chiral edge state, in proximity to a superconductor.

Challenges: A key challenge here is that one must often distinguish a Majorana zero mode from other low-energy vortex-bound states in superconductors. In a large-size superconductor, conventional vortex cores host a series of Caroli-de Gennes-Matricon states, whose lowest-energy state in a fully gapped s-wave superconductor is at $E \approx \Delta^2/E_F$ (typically not exactly zero). In a topological surface, the expectation is that one of these states is pinned exactly to zero by topology, but if the experimental energy resolution is limited, a very low-energy trivial state might be misidentified.

C. Atomic Chains on Superconductors (Shiba Chains)

An alternative approach to engineer a 1D topological superconductor [13] is through chains of magnetic atoms placed on the surface of a superconductor. These are often called Shiba chains, because a single magnetic atom in a superconductor gives rise to Yu-Shiba-Rusinov bound states inside the gap; a chain of many magnetic atoms creates a band of such states, which under certain conditions can become a topological superconducting band. In 2014, Nadj-Perge *et al.* demonstrated this concept by depositing iron (Fe) atoms on the surface of superconducting lead (Pb) and using an STM to assemble and probe Fe chains. The superconducting Pb substrate induces pairing in the electron states of the Fe chain. If the Fe chain is ferromagnetically ordered (which provides an internal exchange field) and if strong spin-orbit coupling from the Pb substrate is present, theory predicts the chain can enter a topological superconducting phase supporting Majorana end modes.

Experimental evidence: The hallmark result from the Fe/Pb chains was the observation of zero-bias peaks at the ends of the chains, but not in the middle, when measured via STM spectroscopy. Nadj-Perge *et al.* found that the differential conductance at the terminus of a chain exhibited a pronounced zero-bias enhancement, whereas spectra taken over atoms in the interior showed a fully gapped superconducting density of states with no zero-bias feature. Moreover, the wave function associated with the zero-bias end-state decayed over only ~10 Å into the chain (just a few atomic spacings), indicating it was strongly localized at the edge. These observations are consistent with the presence of Majorana-bound states localized at the chain ends.

Challenges: There are several limitations to this platform. Firstly, the chains in these experiments are short (on the order of 100 atoms, maybe a few tens of nanometers). If the chain length L is not much longer than the superconducting coherence length ξ, the two end Majoranas will overlap and hybridize, splitting the zero mode into slightly finite-energy modes (since the pair can combine to form a regular fermion). Some theoretical estimates and the experimental data suggest that in the original Fe/Pb chains, $L \ll \xi$ (the coherence length in Pb is relatively long), which is puzzling given the observed strong localization. This discrepancy between expected coherence length and apparent localization length is an ongoing topic of research.

D. Iron-Based Topological Superconductors

Recently, a promising development has been the identification of intrinsic iron-based superconductors that may host Majorana modes without the need for fabricating a heterostructure [15, 16]. The prime example is the iron chalcogenide superconductor $FeTe_{0.5}Se_{0.5}$ (where a portion of tellurium is replaced by selenium). This material is an s-wave superconductor ($T_c \approx 14.5$ K) but, importantly, it has a nontrivial band structure: ARPES measurements showed a band inversion and the presence of a Dirac-cone surface state at the Fermi level. In other words, Fe (Te, Se) is a superconductor that also has topological surface states; it can be thought of as a superconductor with a built-in topological insulator surface. When it becomes superconducting, the surface state acquires a full pairing gap. If that surface state is indeed topologically protected, a magnetic vortex in this superconductor should host a Majorana zero mode.

Experimental evidence: In 2018, several groups reported STM studies of $FeTe_{0.55}Se_{0.45}$ single crystals at high magnetic fields [15-17]. They consistently observed a zero-bias peak in the tunnelling conductance at the center of vortex cores, which in many cases did not split as the STM tip moved around inside the vortex – it remained zero-energy until the edge of the vortex core was reached.

Challenges: Iron-based topological superconductors (Fe(Te, Se)) suffer local in homogeneity and disorder, so vortex spectra can vary and zero-bias features may not be reproducible across all vortices. Trivial vortex-core bound states can also sit near zero energy and mimic Majorana signals unless high-resolution, systematic tests are done. Even if a robust zero mode exists, scalable qubits require controlled creation and motion of vortices and long parity lifetimes against quasiparticle poisoning, and reliable thin-film growth remains challenging, with some leading compounds still relying on tellurium.

Table 1: Summary and Key Characteristics of the Major Platforms Discussed

Platform	*Example Materials (Topological element + Superconductor)*	*Majorana Signatures Observed*	*Notable Challenges*
Semiconductor nanowire	InAs or InSb nanowire + Al or NbTiN superconductor (uses Indium)	Zero-bias conductance peak (sometimes quantized to $2e^2/h$); missing first Shapiro step (4π Josephson effect); hard induced gap.	Disorder-induced trivial zero modes; need magnetic field; uses scarce In metal.
Topological insulator + SC	Bi_2Se_3, Bi_2Te_3 etc. (uses Te) + Nb or $NbSe_2$ superconductor; or magnetic TI (QAH state) + SC	Zero-bias vortex core states in STM; possible half-quantized conductance in edge transport (claimed)	Distinguishing Majorana from other vortex states; material defects; debated experimental reproducibility; Te is extremely rare
Shiba atomic chains	Fe or Co atoms on a superconductor (e.g. Pb, Re) (all elements common)	End-of-chain zero-bias peaks in STM, absent in the middle; strong localization of end state	Short coherence length *vs.* chain length issues; not tunable or easily scalable; operation at very low T to resolve small gap
Iron-based superconductor	$FeTe_{0.5}Se_{0.5}$ (intrinsic SC, topological surface state)	Zero-bias peak in vortex cores that remains unsplit over space; withstands higher T (few K); high quality factors	Requires vortex control, moderate use of Te (rare); need to move or manipulate vortices for qubit operations

III. Societal Impact and Sustainability Considerations

Beyond the physics, Majorana-based quantum computing is pursued for its potential to address the high resource demands of current superconducting systems, which rely on complex refrigeration, extensive control electronics, and large qubit counts. If Majorana topological qubits can be realized, they promise several advantages that could make quantum computing more accessible and sustainable in the long term:

Reduced Error-Correction Overhead: Topological qubits are naturally protected from certain errors, which would drastically cut down the number of physical qubits needed per logical qubit. In today's superconducting qubit schemes, one might need hundreds or thousands of physical qubits to encode a single robust logical qubit with error correction. In contrast, a Majorana qubit might need far fewer, perhaps an order of magnitude or two less overhead. Some projections suggest that the overhead for error correction could be reduced by a factor of ~100 in a topological qubit system. Even Microsoft's recent foray into a prototype topological qubit indicated a potential tenfold overhead reduction for quantum error correction. Fewer qubits mean smaller chip size, fewer control lines, and less complexity in general.

Lower Cryogenic Power Consumption: Operating a quantum computer at millikelvin temperatures is an energy-intensive process. A single dilution refrigerator used for quantum computing can consume on the order of 10-25 kW of power continuously [18], which is equivalent to powering a dozen average homes. This power is mostly used to pump cryogens and run pulse-tube coolers to maintain the cold environment, as well as to power the extensive room-temperature electronics that control the qubits. If we need thousands of qubits for a useful computer, we may require multiple refrigerators or more powerful ones, scaling up the energy usage significantly. Topological qubits promise more practical quantum hardware by reducing qubit count and enabling higher-temperature operation, potentially allowing energy-efficient cryocoolers instead of large dilution refrigerators.

Material Resources and Supply Chain: Majorana devices introduce some new materials compared to conventional silicon or aluminium-based technologies. Indium is heavily used in semiconductor nanowires (InAs, InSb). Indium is not exceedingly rare (~0.05-0.1 ppm in Earth's crust, somewhat more abundant than silver), but it is a byproduct of zinc mining, and its supply is limited and concentrated in certain countries. It's also in high demand for indium tin oxide (ITO) in flat panel displays. If a future quantum computer required millions of InAs nanowires, that could become a significant fraction of indium production. From a device perspective, since topological qubits should need fewer total devices, the material quantity per computation would be small, which helps.

Energy and Sustainability: As quantum computers scale to millions of qubits, their energy and material demands could rival data centres. Topological quantum computing offers a path toward efficiency by reducing qubit counts, simplifying cryogenics, and inherently lowering error-correction overhead-cutting power use and fabrication complexity, and promoting more sustainable, streamlined manufacturing [20, 21].

In sum, Majorana-based quantum computing, if successful, could allow quantum processors that achieve the same computational power with far fewer qubits and less supporting hardware. This directly translates to energy and materials savings: fewer microwave lines, which means less heat load and less copper usage, possibly elimination of some stages of amplification (if topological qubits allow for simpler readout schemes), and crucially, the ability to do more with a single dilution refrigerator.

Finally, from societal perspective, if quantum computers become more efficient thanks to topological qubits, it could democratize access to quantum computing by lowering operational costs. Instead of only tech companies running costly quantum datacenters, smaller institutions could afford to operate quantum hardware if the maintenance and power requirements are reduced. This broader accessibility can accelerate innovation and ensure the benefits of quantum computing (for drug discovery, materials science, optimization, etc.) are widely realized. Thus, quest for Majorana qubits is not only a marvel of fundamental physics but also intertwined with practical goals of sustainability and technological equity.

IV. Conclusion

In this chapter, we have explored Majorana materials: the specially engineered superconductors that can host Majorana zero modes and examined how they might enable more resource-efficient quantum computing. Starting from Kitaev's chain model, we saw how a simple 1D p-wave superconductor supports unpaired Majorana fermions at its ends, providing a template for real-world systems. We surveyed the landscape of experimental platforms: from semiconductor nanowires, which pioneered the search with zero-bias peaks and fractional Josephson effects, to topological insulator hybrids and atomic chains on superconductors, and now to newer high-temperature iron-based superconductors that bring Majorana physics into a more materials-centric realm. Each platform has yielded interesting signatures of Majorana, while also teaching us caution through the lessons of disorder and impostors like trivial bound states.

In conclusion, the pursuit of Majorana-based topological superconductors unites advanced physics with long-term technological vision, aiming to embed resistance to decoherence directly into hardware. Realizing such qubits would mark a major achievement in materials science and quantum engineering: offering robust, low-overhead quantum computation. Despite challenges and mixed progress, the promise of scalable, energy-efficient quantum technology continues to drive the field toward making Majorana qubits a practical reality that bridges science and society.

ACKNOWLEDGEMENT

The author acknowledges the Department of Physics, IIT Tirupati, for the facilities and resources. This work benefited from discussions with colleagues at IIT Tirupati.

REFERENCES

[1] A.Y. Kitaev, Unpaired Majorana fermions in quantum wires, Physics-Uspekhi 44, 131–136 (2001).

[2] A.Y. Kitaev, Fault-tolerant quantum computation by anyons, Annals of Physics 303, 2–30 (2003).

[3] L. Fu and C.L. Kane, Superconducting proximity effect and Majorana fermions at the surface of a topological insulator, Physical Review Letters 100, 096407 (2008).

[4] R.M. Lutchyn, J.D. Sau, and S.D. Sarma, Majorana fermions and a topological phase transition in semiconductor–superconductor heterostructures, Physical Review Letters 105, 077001 (2010).

[5] Y. Oreg, G. Refael, and F. von Oppen, Helical liquids and Majorana bound states in quantum wires, Physical Review Letters 105, 177002 (2010).

[6] J. Alicea *et al.*, Non-Abelian statistics and topological quantum information processing in 1D wire networks, Nature Physics 7, 412–417 (2011).

[7] V. Mourik *et al.*, Signatures of Majorana fermions in hybrid superconductor–semiconductor nanowire devices, Science 336, 1003–1007 (2012).

[8] L.P. Rokhinson, X. Liu, and J.K. Furdyna, The fractional a.c. Josephson effect in a semiconductor–superconductor nanowire as a signature of Majorana particles, Nature Physics 8, 795–799 (2012).

[9] S. Nadj-Perge *et al.*, Observation of Majorana fermions in ferromagnetic atomic chains on a superconductor, Science 346, 602–607 (2014).

[10] J.-P. Xu *et al.*, Experimental detection of a Majorana mode in the core of a magnetic vortex inside a topological insulator–superconductor heterostructure, Physical Review Letters 114, 017001 (2015).

[11] H.-H. Sun *et al.*, Majorana zero mode detected with spin selective Andreev reflection in the vortex of a topological superconductor, Physical Review Letters 116, 257003 (2016).

[12] M.T. Deng *et al.*, Majorana bound state in a coupled quantum-dot hybrid-nanowire system, Science 354, 1557–1562 (2016).

[13] Q. L. He *et al.*, Chiral Majorana fermion modes in a quantum anomalous Hall insulator–superconductor structure, Science 357, 294–299 (2017).

[14] S.M. Albrecht *et al.*, Exponential protection of zero modes in Majorana islands, Nature 531, 206–209 (2016).

[15] D. Wang *et al.*, Evidence for Majorana bound states in an iron-based superconductor, Science 362, 333–336 (2018).

[16] P. Zhang *et al.*, Observation of topological superconductivity on the surface of an iron-based superconductor, Science 360, 182–186 (2018).

[17] H. Zhang *et al.*, Quantized Majorana conductance, Nature 556, 74–79 (2018).

[18] H. Žitko, O. Bodensiek, and T.R. Pruschke, Effects of magnetic impurities on the stability of Majorana fermions in quantum wires, Physical Review B 83, 054512 (2011).

[19] A. Vuik, B. Nijholt, A.R. Akhmerov, and M. Wimmer, Reproducing topological properties with quasi-Majorana states, Science Advances 5, eaaw8873 (2019).

[20] B.J. Kayyalha *et al.*, Absence of evidence for chiral Majorana modes in quantum anomalous Hall–superconductor devices, Science 367, 64–67 (2020).

[21] J.I. Vasquez, J.M.V. Delgado, and D. Serrano, Quantum computing energy consumption: how sustainable is it?, Quantum Reports 3, 80–101 (2021).

Author Index